Reclamation of Mine-Impacted Land for Ecosystem Recovery

Reclamation of Mine-Impacted Land for Ecosystem Recovery

Nimisha Tripathi
Bio-Environment Division
Central Institute of Mining and
Fuel Research

Raj Shekhar Singh
Bio-Environment Division
Central Institute of Mining and
Fuel Research

Colin D. Hills
Faculty of Engineering and Science
University of Greenwich

WILEY Blackwell

This edition first published 2016
© 2016 by John Wiley & Sons, Ltd

Registered Office
John Wiley & Sons, Ltd, The Atrium, Southern Gate, Chichester, West Sussex, PO19 8SQ,
United Kingdom

Editorial Offices
9600 Garsington Road, Oxford, OX4 2DQ, United Kingdom
The Atrium, Southern Gate, Chichester, West Sussex, PO19 8SQ, United Kingdom

For details of our global editorial offices, for customer services and for information about
how to apply for permission to reuse the copyright material in this book please see our
website at www.wiley.com/wiley-blackwell.

Library of Congress Cataloging-in-Publication data applied for

ISBN: 9781119057901

A catalogue record for this book is available from the British Library.

Wiley also publishes its books in a variety of electronic formats. Some content that
appears in print may not be available in electronic books.

Cover image: Courtesy of Author

Set in 10/12.5pt Palatino by SPi Global, Pondicherry, India

Printed and bound in Singapore by Markono Print Media Pte Ltd

1   2016

Contents

Preface

Mining activities significantly impact upon the environment. Across the world, the mining activities generate huge quantities of spoil, promote deforestation and the loss of agricultural production, and the release of contaminants that mean that valuable soil resources are being lost while minerals are being won.

As the effects of the disturbance of ecosystems and loss of valuable land by extractive industries are now being recognized, it is important to show how corrective action can be taken. The introduction of sustainable mining activities does not mean a quantum leap in the technology utilized in mining, but the simple introduction of considered planning and mitigation strategies, that start before mining takes place and extend to after mining has ceased and post-closure activities are being executed.

In this book, the authors have attempted to show how mining impacts on the properties of soil and how soil carbon reserves/soil fertility can be restored when mining has ceased. Restoration involves a coordinated approach that recognizes the importance of key soil properties to enable re-vegetation to take place rapidly and ecosystems to be established in a low cost and sustainable way.

About the authors

Nimisha Tripathi is an Australian Endeavour Fellow, presently working as a visiting academic at the University of Greenwich, UK. Her broad area of research includes restoration and microbial ecology of damaged terrestrial ecosystems. She has worked as a project leader on the rejuvenation of contaminated mine wastelands; has carried out pioneering work on modified chitosan for soil remediation and carbon sequestration. Dr Nimisha has extensive publications in international peer-reviewed journals and a copyright on a novel method developed for estimation of nitrogen in soil and plant materials. She has won several prestigious awards and prizes, including, the Endeavour Research Award (Gov. of Australia), Young Scientist Award (Gov. of India) and, the Green Scientist Award (Hindustan Times and Dainik Jagran Group). Her bio-data is cited in International Biographical Centre, Cambridge, England (2009) and the Marquis 'Who's Who in the World' USA 26th Edition (2009).

Raj Shekhar Singh is Principal Scientist and Associate Professor at CSIR-Central Institute of Mining and Fuel Research, Dhanbad. Dr Raj is specialized in restoration ecology and has extensive research experience on restoration of alternate land uses and damaged ecosystems, remediation of contaminated wastelands, environmental impact assessment and management plan. He has, to his credit, publications in more than 150 journals including *Nature* (London) and books, patents and copyright. Dr. Raj was awarded the UK Commonwealth Fellowship (2012), and had attended the UK House of Parliament to discuss waste recycling (2013) apart from winning several awards and honors for his research contribution, including Whitaker Award (1998), Advisor (Research & Development), Green Earth Citizen, Sweden (2013) and CSIR patent prizes (2006, 2007). His Bio-data was cited in the Dictionary of International Biography Centre, Cambridge, England (1997). Dr Raj has edited more than 10 proceedings and is the advisor and editorial board member of a number of peer-reviewed journals.

Colin D. Hills is Professor of Environment and Materials Engineering at the University of Greenwich. Professor Hills has an extensive research and publishing record on the treatment and valorisation of hazardous wastes and contaminated soils. He has authored national guidance on stabilisation/solidification technology for the Environment Agency, is Academic Lead for the CO_2-Chemistry KTN (Utilisation Cluster), is European Contributor to the UN EP Global Environment Outlook (GEO6) for Waste and Chemicals and is a contributor to the EU road map for CO_2 mineralisation. Professor Hills has won a number of major awards for his work, including: the Green Chemical Technology Prize (IChemE), National Winner of the Shell Springboard Challenge (2008), winner of the Times Higher Award for his Outstanding Contribution to Innovation and Technology (2008). Professor Hills is Founder Director, Technical Director of Carbon8 Systems Ltd, and a Founder of Carbon8 Aggregates Ltd, sister companies that are pioneering the use of waste CO_2 gas for the engineering of waste materials.

Acknowledgements

The authors wish to acknowledge the invaluable help received from a number of established researchers on mine-impacted lands at the Bio-Environment Division of Central Institute of Mining and Fuel Research (CIMFR), Dhanbad. Completion of this book would not have been possible without their encouragement. The Director, CIMFR, Dhanbad, India, is gratefully acknowledged as is Prof. J.S. Singh, Emeritus Professor, BHU, Varanasi, for their unselfish support. Of particular note are the contributions made by Dr C.S. Jha, Scientist, NRSC, Hyderabad, Prof. A.S. Raghubanshi, BHU, Varanasi, Dr S.C. Garkoti, Associate Professor, JNU, New Delhi and Dr S.K. Chaulya, Scientist, CIMFR, Dhanbad.

The authors would also like to thank the Department of Science and Technology, Government of India, New Delhi, for research funding support on the rejuvenation of contaminated mine wastelands and the Ministry of Coal, Ministry of Rural Development and Employment, for also providing financial support on a number of research initiatives in this important area of environmental impact management.

Finally, we are indebted to our family members who have provided unselfish support during the writing of this book. Dr Raj S. Singh wishes to express much gratitude to his late father, and Dr Nimisha Tripathi to her late grandfather for his inspiration throughout this endeavour.

1 Introduction

1.1 Background and purpose

Since Palaeolithic times (ca. 450000 years ago), mining has been an integral part of the human existence (Hartman, 1987). Mining is fundamental to technological development and there is evidence of subsurface mining dating back to 15000 BC (Kennedy, 1990).

Throughout the world, the most common form of mineral extraction is surface or open-pit mining. Minerals with a low stripping ratio generate large amounts of overburden or spoil, which are discarded on adjacent land surface.

The discarded overburden is disposed of in surface dumps, which significantly impact upon both flora and fauna. Spoil dumps occupy large areas of productive land and contaminate surface and subsurface water resources, upon impacting ecological pools and biological processes (Tripathi et al., 2012). The loss of key components of an ecosystem directly results in land degradation.

Surface mining disrupts the environment by disturbing the landscape, despoiling agricultural land and through deforestation. The consequence of mining is a loss of plant biomass and land productivity. The environmental impacts caused by mining, based on Richards (2002), are:

- Ecosystem disturbance and degradation
- Habitat destruction
- Adverse chemical impacts (from improperly treated wastes); and
- Loss of soil-bound carbon (to the atmosphere)

The management of mine spoil/degraded land is a major issue throughout the world. The ecological and environmental impacts of mining warrant a corrective action supported by appropriate post-closure

Reclamation of Mine-Impacted Land for Ecosystem Recovery, First Edition. Nimisha Tripathi, Raj Shekhar Singh and Colin D. Hills.
© 2016 John Wiley & Sons, Ltd. Published 2016 by John Wiley & Sons, Ltd.

management strategies. By managing environmental impacts, the long-term viability of mining operations can be secured.

The practice of ecological restoration of disturbed and degraded land is a primary action in ecosystem recovery. This is achieved by ensuring a nutrient cycling is re-established, which in turn fosters increasing biodiversity.

The introduction of a progressive post-mining plan, which considers the ecological condition of the land (to be mined) and the suitability of native plants for reclamation activities is an important step as this:

- Minimizes the overall impact of mining at a site
- Ensures an appropriate post-mining closure design is implemented
- Reduces overall cost
- Enhances environmental protection and restoration of soil-based carbon
- Reduces the time frame for completing the reclamation strategy

Post-closure reclamation actions can be implemented immediately after the cessation of mining and should utilize the best available technology options available.

Thus, by using appropriate management strategies, such as mulches and organic matter-based additions, re-vegetation can be effectively carried out post mine closure. Reclamation will re-establish the soil carbon reserve lost during mining that is essential for the correct functioning of vegetation. The reintroduction of soil organic matter is achieved via the removal of CO_2 from the atmosphere into root mass and leaf litter. The growth of biomass reduces the amount of CO_2 in the atmosphere, and therefore mitigates the effects of climate change.

This work provides a comprehensive description of impacts arising from land degradation caused by mining activities. It provides insight into the technical aspects of the restoration and reclamation of mining-impacted land and the reintroduction of soil-based carbon reserves that are so important to the re-establishment of self-sustaining ecosystems. Key ecological concepts are explored, and the major ecological pools and biological processes functioning in disturbed or degraded ecosystems are presented.

The successful repair of degraded land and reintroduction of a sustainable ecosystem requires a multidisciplinary approach, and this is reflected in the content of this book. All the stages of land reclamation from the initial policy decisions to management and outcomes are presented. As such, this work will provide key insights to undergraduate and postgraduate students, researchers, mine managers, policymakers and professionals dealing with contaminated mine land reclamation and management issues.

1.2   Key concepts and definitions

A number of key concepts and definitions are presented which underpin the understanding of ecological restoration. A number of these are as follows:

Biogeochemical cycles	The pathway by which a chemical substance moves through both biotic (biosphere) and abiotic (lithosphere, atmosphere and hydrosphere) components of Earth.
Carbon sequestration	This is the process of naturally or artificially storing carbon dioxide for a longer-term out of the atmosphere, where it contributes to the greenhouse effect.
Carbon sink	A natural or artificial reservoir that accumulates and stores carbon-containing chemical compounds (e.g. CO_2) for an indefinite period.
Decomposition	Conversion or decay of chemically unstable material to simpler forms by the natural action of air, water, light and microorganisms.
Disturbance	The major cause of long-term changes in the structure and functioning of ecosystems. Disturbance may be natural, involving fire, wind, disease, insect outbreaks and landslides, or anthropogenic from human impacts (e.g. clear cutting, deforestation, habitat destruction, introduction of invasive species).
Ecology	A branch of biology dealing with the interactions among organisms and their abiotic environment: the study of 'the structure and function of nature, which includes the living world' (Odum, 1959). In terms of disturbance, ecology encompasses the study of interrelationships between biotic and abiotic components of the existing disturbed ecosystems.
Ecosystem	A biological community of interacting organisms and their physical environment. Ecosystems are characterized as complex systems with abiotic and biotic processes interacting between the various components. In simple terms, Odum's ecosystem is the fundamental unit of ecology.
Ecological processes	The key processes regulating the ecological system (ecosystem) – nutrient processing, productivity, decomposition, nutrient turnover, hydrological flux.
Ecological restoration	The practice of renewing and returning a degraded, damaged or destroyed ecosystems to its original (prior to disturbance) condition.
Ecosystem development	The development of pools and processes of an ecosystem culminating in a stabilized ecosystem. Ecosystem development is the part of ecological succession. The concepts of ecosystem development are often based on assumptions and extrapolations with respect to structural–functional interactions in the initial stage of ecosystem development.

(continued)

(*Continued*)

Ecosystem productivity	In ecology, productivity refers to the rate of generation of biomass in an ecosystem.
Endemic (or native) plants	The plant species indigenous and unique to a specific geographic region over a given period of time.
Exotic plants	The plant species living outside its native distributional range, which has arrived there either by deliberate or accidental human activity.
Functional components	The components of ecosystem having specific roles in regulating the functioning (e.g. biogeochemical processes, disturbance regimes) of an ecosystem but governed by the structural components. Four functional components of an ecosystem include:

- Abiotic factors
- Producers
- Consumers
- Decomposers

Odum (1959) termed the three 'functional kingdoms of nature' for latter three living components.

Greenhouse effect	The phenomenon by which the sun's thermal radiation is trapped by the gases (e.g. carbon dioxide, methane, water vapour) of a planetary surface and is re-radiated back from the planet causing atmospheric heating.
Greenhouse emission	The emission of gases, for example, chlorofluorocarbon, carbon dioxide, perfluorocarbon, sulphur hexafluoride, that contributes to the greenhouse effect by absorbing infrared radiation.
Habitat alteration	The process making changes to the environment that adversely affects ecosystem function. However, the effects are not permanent (Dodd and Smith, 2003).
Habitat destruction	The process in which natural habitat is rendered functionally unable to support the existing species. In this process, the regional ecosystem is completely eliminated resulting into the total removal of its former biological function and loss of biodiversity (Dodd and Smith, 2003). Habitat destruction is the primary cause of species extinction worldwide.
Habitat fragmentation	A secondary affect of habitat destruction, which occurs when the remaining species populations after habitat destruction are isolated due to destroyed linkages between habitat patches after disturbance.
Land disturbance	Changes of land use and land forms, soil moisture regulation, loss of biodiversity, loss of soil organic matter pool and altered nutrient cycling.
Land degradation	As defined by the UN Environment Programme, land degradation is 'a long-term loss of ecosystem functions and services, caused by disturbances from which the system cannot recover unaided' (Dent, 2007).
Land reclamation	The act of returning a land to a former, better state. In terms of a wasteland, land reclamation refers to the conversion of wasteland into useful land.
Land rehabilitation	The act of returning a damaged land to some degree of its former state.

(*continued*)

(*Continued*)

Litter	Fallen leaves and other decaying organic matter that make up the top layer of a terrestrial ecosystem.
Mine spoil/overburden	Miner's definition: Any loose or consolidated material lying over a mineral deposit of ore or coal. Civil engineer or soil scientist definition: loose soil, sand, or gravel lying above the bedrock.
Natural disturbance regime	This is a concept that describes the pattern of disturbances that shape an ecosystem over a long timescale. It describes a spatial disturbance pattern, a frequency and intensity of disturbances, and a resulting ecological pattern over space and time. These disturbances do not include the anthropogenic disturbances.
Resilience	The capacity of an ecosystem to respond to a disturbance by resisting damage and recovering quickly.
Standing biomass	The total dried biomass of the living organisms present in a given environment.
Soil amendments	Soil amendments are the materials added to soil to improve the quality of soil, especially its ability to provide nutrients to plants. They also act as the soil conditioners.
Soil microbes	Microorganisms for which the soil is the natural habitat. Examples include bacteria, actinomycetes, fungi, algae and protozoa.
Stripping ratio	The unit amount of overburden that needs to be removed to access/extract a similar unit of coal, mineral/metal ore.
Structural components	The structural components of an ecosystem are constituted by living (biotic) and nonliving (abiotic) components. Living components: populations of organisms (species diversity) and the living resources they use. Nonliving components: nonliving resources (e.g. space) and the nonliving physical characteristics of habitats (e.g. temperature, humidity, habitat complexity).
Succession	The process by which the structure of a biological community evolves over time.
Sustainable development	Development that meets the needs of the present without compromising the ability of future generations to meet their own needs. Sustainable development promotes the idea that social, environmental and economic progresses are all attainable within the limits of earth's natural resources.

1.3 Supporting information

The mitigation of environmental impacts from mining activities is a complex subject. In order for the reader to access more information on particular aspects of mine restoration, a number of information sources are given below in Tables 1.1, 1.2 and 1.3. These include the addresses of organizations involved in mine restoration in India and elsewhere, useful websites and a list of NGOs involved with restoration activities (Tables 1.4 and 1.5).

Table 1.1 List of relevant organizations.

India

Banaras Hindu University (BHU), Varanasi
Central Institute of Mining and Fuel Research (CIMFR), Dhanbad
Central Soil and Water Conservation Research and Training Institute (CSWCRTI), Dehradun
Forest Research Institute (FRI), Dehradun
Indian School of Mines (ISM), Dhanbad
National Environmental Engineering Research Institute (NEERI), Nagpur
Tropical Forest Research Institute (TFRI), Jabalpur

International

International Affiliation of Land Reclamationists (IALR): an umbrella organization,
    which encompasses restoration groups in the United Kingdom, the United States,
    Australia, Canada and China:
- The British Land Reclamation Society
- American Society for Surface Mining and Reclamation
- Mineral Council of Australia
- Canadian Land Reclamation Association (Association Canadienne de Rehabilitation
    des Sites Degrades)
- China Land Reclamation Society

UK Environment Agency (EA)
American Society of Mining and Reclamation (ASMR), Virginia
Intergovernmental Panel on Climate Change (IPCC), Switzerland
International Union for Conservation of Nature (IUCN), Switzerland
Interstate Mining Compact Commission (IMCC), New York
National Association of State Land Reclamationists (NASLR), New York
Office of Surface Mining, Reclamation and Enforcement (OSMRE), New York
United Nations Environment Program (UNEP), Geneva
The United Nations Educational, Scientific and Cultural Organization (UNESCO), Paris
Western Pennsylvania Coalition for Abandoned Mine Reclamation, Pennsylvania
World Wide Fund for Nature (WWF), Switzerland

1.4   Structure/layout of the book

To address the impact of surface mining on terrestrial ecosystem and its management, the authors have used India as the prime model while discussing the general reclamation practices and underlying policies worldwide.

The book is organized in seven chapters to cover the ecological principles of land restoration, adequate management of degraded mine lands and the consequent environmental and societal benefits. Included is a synthesis of the authors experience for more than 20 years of research on mine-degraded lands and their environmental impact and subsequent reclamation.

Table 1.2 List of NGOs involved in eco-restoration.

NGO	Activities
India	
Foundation for Ecological Research, Advocacy and Learning, Pondicherry	Wildlife conservation, ecological restoration, natural resource management and capacity building
Foundation for Ecological Security (FES), Anand (Gujarat)	Ecological restoration and conservation of land and water resources in ecologically fragile, degraded and marginalized regions of the country through collective efforts of village communities
Centre for Science and Environment (CSE), Delhi	As a think tank for environment–development issues, poor planning, climate shifts, devastating India's Sundarbans, policy changes and better implementation of the already existing policies
Dasholi Gram Swarajya Mandal, Gopeshwar, Chamoli	Forest conservation and eco-regeneration; use of forest products for self-employment
Green Future Foundation, Pune, Maharashtra	Environmental protection, energy and ecological conservation and pollution control
Rajasthan Environment Preservation Society, Jaipur	Pollution control, afforestation, ecological and environmental preservation
The Energy and Resources Institute (TERI), Delhi	Policy-related work in the energy sector, increased biomass production, conversion of waste into useful products and mitigating the harmful environmental impacts of several economic activities
United Kingdom	
British Land Reclamation Society	Reclamation, rehabilitation and restoration of contaminated, derelict and abandoned mine and industrial land
Contaminated Land: Applications in Real Environments (CL: AIRE)	Regeneration of contaminated land for sustainable remediation
Flora locale	Conservation and enhancement of native wild plant populations and plant communities in the context of ecological restoration and creative conservation
The Land Trust	The transformation of land unsuitable for development into high-quality public open space (such as country parks, wetlands, community woodlands and ecology parks)
Groundwork	Environmental regeneration, prevention of underuse of land and exploring appropriate use of land
Waste Watch	Sustainability, environmental protection/restoration, well-being
United States	
The Society for Ecological Restoration (SER)	Promoting ecological restoration and conservation
World Association of Soil and Water Conservation, Iowa	Solving scientific and technical problems related to soil and water conservation
Soil and Water Conservation Society (SWCS)	Foster the science and art of natural resource conservation
Air and Waste Management Association, Pennsylvania	Environmental management, critical environmental decision-making

(*continued*)

Table 1.2 (*Continued*)

NGO	Activities
Denmark	
COWI	Land restoration, environmental impact assessment
The Netherlands	
International Institute of Land Reclamation and Improvement (ILRI)	Sustainable use of land and water resources, especially in developing countries
International Association	
International Erosion Control Association	Erosion and by-product–sediment control
• Region one: North America, South America and Europe	
• Region two: Australasia, Asia and Africa	

Table 1.3 List of abbreviations.

C	Carbon
CO_2	Carbon dioxide
CEC	Cation exchange capacity
CIL	Coal India Ltd
CPCB	Central Pollution Control Board
EA	Environmental assessment
EMP	Environmental management plan
EU	European Union
GDP	Gross domestic product
GHG	Greenhouse gas
HZL	Hindustan Zinc Ltd
IEA	International Energy Agency
IPCC	Intergovernmental Panel on Climate Change
IUCN	International Union for Conservation of Nature
MSW	Municipal solid wastes
NPK	Nitrogen, phosphorus and potassium
NPP	Net primary productivity
N	Nitrogen
OECD	Organisation for Economic Co-operation and Development
OM	Organic matter
OSM	Office of Surface Mining
PPP	Public–Private Partnership
R&R	Resettlement and rehabilitation
SER	Society for Ecological Restoration
SOC	Soil organic carbon
SOM	Soil organic matter
SAIL	Steel Authority of India Ltd
SMCRA	Surface Mining Control and Reclamation Act
UNEP	United Nations Environment Program
UNESCO	United Nations Educational, Scientific and Cultural Organization
WASCOB	Water and sediment control basin
WHC	Water-holding capacity
WEO	World Energy Outlook
WTE	Waste-to-energy

Table 1.4 List of key reference sources (website links).

http://blogs.scientificamerican.com/observations/2013/05/09/
  400-ppm-carbon-dioxide-in-the-atmosphere-reaches-prehistoric-levels/
http://envfor.nic.in
http://india.indymedia.org/en/2002/12/2456.shtml
http://moef.nic.in/downloads/home/home-SoE-Report-2009.pdf
http://www.cseindia.org/programme/industry/mining/political_minerals_mapdescription.htm
http://www.nrsc.gov.in/pdf/P2P_JAN11.pdf
http://scclmines.com/scclnew/careers/docs/Notification012015.pdf
http://scclmines.com/downloads/exploration.pdf
http://timesofindia.indiatimes.com/india/New-land-acquisition-law-comes-into-force/articleshow/
  28204302.cms
http://www.waste-management-world.com/articles/2003/07/an-overview-of-the-global-waste-to-
  energy-industry.html
http://www.cci.in/pdfs/surveys-reports/Mineral-and-Mining-Industry-in-India.pdf
http://www.cseindia.org/node/386
http://www.eldoradochemical.com/fertiliz1.htm
http://www.globalrestorationnetwork.org/restoration/methods-techniques/
http://www.iea.org/textbase/nppdf/free/2000/coalindia2002.pdf
http://www.moef.gov.in
http://envfor.nic.in
http://www.rediff.com/money/2003/aug/15waste.htm

Table 1.5 List of units used.

Bn	Billion
cm	Centimetre
Gt	Gigatonnes
g	Gram
ha	Hectare
kg	Kilogram
km	Kilometre
Mg	Million gram
Mha	Million hectare
Mpa	Megapascal
Mt	Million tonnes
MT	Metric tonnes
Pg	Peta gram
Ppm	Parts per million
te	Tonne
Tg	Teragram
μm	Micrometre

2 Mining and ecological degradation

2.1 Background

Currently human activities are impacting negatively upon the health of ecosystems at unprecedented rates worldwide. The generation of huge amounts of waste material causes significant environmental harm and accelerates the loss of biodiversity.

Modern development relies heavily on the mining industry to maintain the supply of materials to industry. During surface mining, 2–11 times more land is damaged than with underground mining. The direct effects of surface mining activities can be unsightly, but more seriously, these are responsible for the loss of cultivated/arable land, pasture and native forest. The indirect effects can be equally as serious as they include soil erosion, air pollution, pollution of surface and underground water resources, the loss of genetic biodiversity and the impairment of the local economic and general well-being.

In describing the effects of mining and how these can be mitigated, the authors have used India as their prime model. India is the seventh largest and second most populous country, containing 17% of the world's population, while covering 2.5% of the surface of the world. India is 3214 × 2993 km in size and has a land frontier of 15 200 km and a coastline of 7517 km. In India, 306 Mha are associated with mining activities, and of this 147 Mha can be considered as degraded land.

The population of India is currently 1.2 Bn, rising to 1.6 Bn by 2050 (*State of Environment Report*, 2009; http://www.moef.gov.in, http://envfor.nic.in). The rapid industrialization of India and the associated exploitation of natural resources have rapidly accelerated the degradation of natural habitats (Bradshaw, 1983).

However, degradation is not synonymous with disturbance; disturbance becomes degradation when the natural resilience of an ecosystem becomes exhausted. Land degradation is characterized by a loss of

Reclamation of Mine-Impacted Land for Ecosystem Recovery, First Edition. Nimisha Tripathi, Raj Shekhar Singh and Colin D. Hills.
© 2016 John Wiley & Sons, Ltd. Published 2016 by John Wiley & Sons, Ltd.

productivity through negative impacts, including erosion, salt ingress, waterlogging, nutrient depletion and the deterioration of soil structure. Land degradation occurs due to anthropogenic pressures such as deforestation, mineral extraction, industrialization, overgrazing, waste disposal and the excessive use of pesticides. Wind erosion and waterlogging of soils are priority areas of concern.

2.1.1   The need for land reclamation

The practice of land restoration is a primary option for increasing levels of biodiversity by modifying human-altered ecosystems (Brudvig, 2011). The Committee of the National Academy of Science, USA, defined restoration as:

> The replication of site conditions prior to disturbance; whereas the term reclamation refers to rendering a site habitable to indigenous organisms; and rehabilitation implies that disturbed land will be returned to a form and productivity in conformity with a prior land use plan including stable ecological state that does not contribute substantially to environmental deterioration and is consistent with surrounding aesthetic values. (National Academy of Science, 1974)

The degradation of an ecosystem is caused by the loss of key 'components' of the ecosystem. Therefore, in some cases, restoration, in the strict sense, may be impossible. Cairns (1986) pointed out three options for restoration:

(1)   Full restoration: restoration of a site to its pre-damaged condition
(2)   Partial restoration: restoration of selected ecological attributes of the site
(3)   Creation of an alternative ecosystem type (although not strictly restoration)

Mining is a destructive activity and requires careful management and regulation to reduce environmental harm. In developing countries, however, the management of adverse impacts can take second place to the need for economic growth. Thus, India is a perfect case example of the impacts of mining on ecosystems as this extractive activity is widespread throughout the country and environmental harm is widely documented.

2.2   Mining in India

India is rich in natural resources, including coal, iron ore, manganese, bauxite, chromite, diamonds and limestone.

India relies heavily on the mining industry for its basic raw material needs, and its mining industry is the second largest after agriculture. Mining is one of the largest employers and accounts for about 2.3% of the total GDP.

The number of working mines (excluding minor minerals, petroleum (crude), natural gas and atomic minerals) was 2076 in 2011–2012. Of these reporting mines (defined as 'a mine reporting production or reporting "Nil" production during a year but engaged in developmental work such as overburden removal, underground driving, winzing, sinking work; exploration by pitting, trenching or drilling as evident from the MCDR Returns'), 354 were located in Andhra Pradesh followed by Gujarat (308), Rajasthan (241), Madhya Pradesh (225), Karnataka (180), Tamil Nadu (156), Odisha (119), Jharkhand (106), Chhattisgarh (99), Maharashtra (86) and Goa (70). These 11 states together accounted for 93.64% of total number of mines in the country in that year (Ministry of Mines, 2012).

India produces 89 minerals including (Ministry of Mines, Government of India, 2014):

- 4 types of fuel
- 10 metallic minerals
- 48 non-metallic minerals
- 3 atomic minerals
- 24 minor minerals (including construction materials)

The 'value distribution' shows that fuels accounted for about 66%, metallic minerals about 20%, non-metallic minerals about 2% and minor minerals about 12%. Of the metallic minerals, iron ore accounted for about 83% of the 'value' created (*Indian Minerals Yearbook*, 2011).

Coal mining accounts for 80% of mining activity, with the remainder involving gold, copper, iron, lead, bauxite, zinc and uranium (www.cci.in/pdfs/surveys-reports/Mineral-and-Mining-Industry-in-India.pdf). Mining is a constantly growing business with a current production base of 204.95 points (2010–2011) relative to the 1993–1994 (a baseline of 100). The total value of mineral production (including minor minerals but excluding atomic minerals) increased 17% in 2010–2011 over the previous year to Rs. 232021 2 (US$$3.74 \times 10^{11}$).

Several Indian states including Rajasthan, Chhattisgarh, Orissa and Andhra Pradesh are rich in minerals, especially non-ferrous and ferrous metals/minerals. There are considerable deposits of lignite, bituminous and sub-bituminous coals, iron ore and bauxite. A list of different minerals found in different geographical areas of India is given in Table 2.1.

Mining operations necessitate deforestation, habitat destruction, biodiversity erosion and destruction of geological records, which contain

Table 2.1 Minerals in different geographical locations in India.

North Eastern Peninsular Belt	The region comprising the Chota Nagpur plateau and the Orissa plateau which covers the states of Jharkhand, West Bengal and Orissa	Manganese, bauxite, copper, coal, iron ore, mica, kyanite, chromite, beryl, apatite, etc. India's 100% kyanite, this region accounts for the country's 93% iron ore production and 84% coal production
South Western Belt	Karnataka and Goa	Garnet iron ore and clay
North Western Belt	Rajasthan and Gujarat along the Aravali Range	Mostly non-ferrous minerals, uranium, aquamarine, petroleum, mica, beryllium, gypsum and emerald
Southern Belt	Karnataka plateau and Tamil Nadu	Bauxite and ferrous minerals
Central Belt	Andhra Pradesh, Chhattisgarh, Madhya Pradesh and Maharashtra	Bauxite, uranium, manganese, limestone, mica, graphite, marble, coal, gems, etc.

Source: India Mineral Map. http://www.mapsofindia.com/maps/minerals/, accessed 12 June 2015.

information about past biodiversity. Mining is a primary activity generating overburden waste, tailings, and leachate/mine-water seepage. Surface mining operations destroy flora and fauna and contaminate soil, air and water.

After the extraction, an ore is beneficated by crushing and grinding, followed by concentration to separate the valuable mineral fraction from the host rock. Beneficiation involves physical/chemical separation techniques, such as gravity concentration, magnetic separation, electrostatic separation, flotation, solvent extraction, electro-winning, leaching, precipitation and amalgamation.

The beneficiation processes generates tailings normally as a slurry of 30–60% solids. Most mine tailings are disposed of in on-site impoundments/ponds. According to a State of Environment Report (2009), mining sites currently occupy around 0.06% of the total land area of India (http://moef.nic.in/downloads/home/home-SoE-report-2009.pdf), and this area will significantly increase with time.

2.2.1　Coal

Globally, coal is one of the most abundant and important primary sources of energy and plays a crucial role in the world energy production. Reserves of coal are spread worldwide throughout some 100 developed and developing countries, sufficient to meet global needs for the next 250 years. Coal is an important fossil fuel but is also used for other industrial purposes.

India has the world's third largest hard coal reserves after the United States and China with an output of 328 Mt in 2001–2002. It remains the primary source of energy, accounting for about 80% of total energy generation in India. Coal contributes about 26% of the total global primary energy demand and is also a key input for the steel and other industries (http://scclmines.com/downloads/exploration.pdf). Some 70% of India's domestic coal production is used for internal power generation, with the remainder being used by heavy industry and public use. According to the 2008 BP Statistical Energy Survey (2007), India had consumed 208 Mt of coal (oil equivalent). Although most of India's coal has a high ash content, it is generally low in sulphur (≤0.5%), iron and chlorine and has a low refractory nature.

About 88% of the total coal is produced by various subsidiaries (390 mines) of Coal India Ltd. (CIL), the largest supplier of coal (and one of the largest taxpayers) in India.

Coal India is currently state controlled and has seven coal-producing subsidiary companies, namely, Central Coalfields, Eastern Coalfields (Sanctoria), Bharat Coking Coal (Dhanbad), Northern Coalfields (Nagpur), Western Coalfields, Southern Eastern Coalfields (Bilaspur), Mahanandi Coalfields (Sambalpur, Orissa) and the Central Mine Planning & Design Institute (CMPDI) at Ranchi (Jharkhand), which are entrusted with the job of providing total research and consultancy support to the industry.

Private mines in India are allowed to operate only when supplying a specific industry. The only other major (and private) producer is the Singareni Collieries Company Ltd. (SCCL) that is located in Andhra Pradesh. Singareni has 37 underground and 13 opencast mines producing 40.6 Mt of coal in 2007, rising to 52.2 Mt in 2011–2012. The Company proposed to open nine new mines based around new cutting technologies and High Wall Mining in opencast projects (http://scclmines.com/careers/docs/Notification.pdf).

The production of coal has increased from 211 Mt in 1990–1991 to 407 Mt in 2005–2006, with opencast mining contributing 252 Mt (www.cci.in/pdfs/surveys-reports/Mineral-and-Mining-Industry-in-India.pdf). According to the BP Statistical Energy Survey (2010), India had coal reserves of 58.6 Gt (>7% of the world total reserves) (*Coal Mining in India-Overview*, 2005; www.mbendi.com/indy/ming/coal/as/in/p0005.htm).

Traditionally, coal mining is considered to be one of the most polluting industries, with the mining of reserves either by opencast or underground process having significant environmental impacts if proper management strategies are not adopted.

2.2.2 Iron ore

Iron ore mines are found in at least 50 countries around the world, with the United States, Canada, Australia, China, Brazil and India being the main producers. The iron ore deposits comprising hematite, taconite, goethite and magnetite are the most common, with hematite providing the highest iron yield.

There are no mines on Earth that have pure iron deposits, and the stripping ratio ranges from 2 to 2.5 (BBY Limited Resources, POSCO, Murchison Metal Limited), meaning that for every tonne of iron ore produced, double the quantity of waste is generated. In 2003–2004, the Steel Authority of India Ltd. (SAIL) generated 4.8 Mt of overburden and rejects from its 12 mines in the country (Environmental Performance Report of SAIL).

In India, iron ore production has increased almost threefold from 1990 to 2006 (55.5–154.4 Mt, respectively) (http://www.cci.in/pdf/surveys_reports/mineral-mining-industry.pdf). In India, iron ore production was increased almost threefold from 55.5 Mt in 1990–1991 to 154.4 Mt in 2005–2006 (http://www.cci.in/pdf/surveys_reports/mineral-mining-industry.pdf); however, the production was declined by 18.6% from 2010–2011 to 2011–2012 and by 19.3% from 2011–2012 to 2012–2013 (Ministry of Mines, 2012). On average, however, 2.5–3 te of mining waste is excavated for every 1 te of iron ore (http://www.cseindia.org/node/386). As with other extracted ores, the basic steps in mining iron ore are:

- Exploration
- Mine development
- Extraction operation
- Beneficiation of ore
- Storage and transport of ore
- Mine closure and reclamation

The development of a mine involves establishing extraction facilities containing water supply/effluent and tailing processing and treatment/disposal facilities. The closure of mine/mine workings involves reclaiming/stabilizing spoil/waste dumps and introducing measures to stop soil erosion and to return the land to the condition experienced prior to mining.

Waste rock and soil are often used for site reclamation, and soil erosion is avoided by grading slopes followed by seeding of grasses and shrubs. However, the most important step as far as environmental issues are concerned is the plantation of appropriate tree species to maintain the ecosystem balance and conserve the biodiversity of the affected area.

2.2.3   Copper, lead and zinc mining

Copper (Cu), lead (Pb) and zinc (Zn) are extensively mined and generate waste that is toxic to plants. However, there is no estimate of how much Cu waste is generated globally. The production of 1 te of copper generates 110 te of waste ore and 200 te of overburden (Anon, 2006). Thus, globally, the copper industry in 2004 may have generated 3.3 Gt waste materials to yield 10.8 Mt of copper metal (Taylor, 2006).

India has limited copper ore deposits comprising about 2% of the worlds reserves, but with a very low reserve to production and reserve to resource ratio. The mining of copper in India contributes ca. 0.2% of world's production, whereas copper refining generates 4% of world's production (Hindustan Copper Ltd.; http://mines.nic.in/). The production of copper concentrate in 2010–2011 was 137 Mt, and this represented an increase of >9%, compared to 2009–2010. In India, copper mines are located in Madhya Pradesh, Rajasthan, Bihar and Sikkim and copper is closely mixed with sulphur and iron to form the sulphide minerals, chalcopyrite ($CuFeS_2$), bornite (Cu_5FeS_4), covellite (CuS) and chalcocite (Cu_2S). Almost half of the total amount of Cu is obtained from chalcopyrite, with the average metal content in India ore being 1% w/w, relative to a world average of 1.5% w/w.

India has estimated copper reserves of 730 Mt, averaging 1.17% w/w copper, but copper ore production has been sporadic with a fall being recorded between 1991 and 2006 (5.3–2.6 Mt, respectively), whereas in 2009, production was up by 3.5% (http://www.cci.in/pdf/surveys_reports/mineral-mining-industry.pdf).

The production of lead and zinc ore was 8.6 Mt in 2012–2013, a increase of 7% over the previous year (IBM report, Government of India; http://ibm.nic.in/writereaddata/files/10202014112509msmpmar14_07_Highlights_Eng.pdf). In India, the major two organizations, Hindustan Zinc Ltd. (HZL) and Indian Lead Ltd., produce this metal. The production of zinc by HZL is 65 kt/Pa, and 24 kt/Pa for Indian Lead Ltd. Worldwide, however, 60% of the lead produced is from secondary sources (Pappu et al., 2007). Lead-acid batteries are the dominant source of this secondary lead, but the metal also comes from scrap, sheets and pipes, cable sheathing and solder in addition to dust and dross.

Zinc-lead smelting generates annually about 142 kt/Pa of slag as solid waste that require managing (Hindustan Zinc Ltd. 15 kte/Pa; Tundoo lead smelter 18 kte/Pa; HZL Chanderiya 116 kt/Pa and Indian Lead Ltd., Thane 3 kt/Pa). Some of these wastes are recycled for silver, lead and copper recovery, and the rest are disposed to secure landfill (Pappu et al., 2007). During 2010–2011, the production of lead concentrates was 145 kt, representing an increase of 8.3% over the previous year (Ministry of Mines, 2012).

Presently, >75% of the world's zinc is produced via the acid leaching of jarosite, producing large quantities of mud that are stored in closed containers or sealed reservoirs due to the presence of toxic species including Zn, Pb, Cd and S.

In India, HZL has a mining/smelting facility with an installed capacity of 3.5 Mt year^{-1}. Zinc is manufactured from four smelters located in Rajasthan, Andhra Pradesh, Bihar and Orissa. The Debari Zinc Smelter, Rajasthan, is one of India's largest facilities and uses an ammonium jarosite electrolytic extraction method to produce 59 kte year^{-1} of zinc metal. Other major producers of jarosite include China, the United States, Spain, Holland, Canada, France and Australia (Pappu et al., 2007).

During the year 2010–2011, the production of zinc concentrate in India was >1420 kte representing an increase of 11% over the previous years' production (Ministry of Mines, 2012).

2.2.4   Bauxite mining

India is the fifth largest producer of bauxite after Australia, Guinea, Brazil and Jamaica with substantial high grade reserves, estimated at 2600 Mt and expected to last over 350 years at an anticipated consumption of 7 Mt year^{-1} (http://www.mbendi.com/indy/ming/baux/as/in/p0005.htm#5). The production of bauxite at 16 Mt during 2012–2013 was increased by 13% as compared to the previous year, with the highest contribution by Orissa (36%) followed by Gujarat (20%), Jharkhand and Maharashtra (13% each), Chhattisgarh (12%) and Madhya Pradesh (5%). The remaining 1% was contributed by Goa, Karnataka and Tamil Nadu (IBM report, Government of India; http://ibm.gov.in/msmpmar13_07_Highlights_Eng.pdf).

The extraction of bauxite by strip-mining involves the removal all native vegetation in the surrounding area, leading to widespread habitat destruction and soil erosion. The solid by-product from the extraction of alumina from bauxite (the Bayer process) is red mud. Approximately 1.0–2.5 t is produced in India for every 1.0 t of alumina produced (Paramguru et al., 2005). Although red mud presents significant environmental issues (contains heavy metals, is alkaline and is stored in lagoons/ponds), the Indian Bureau of Mines (1977), estimate that from 1 Mt of red mud, about 300 kt of iron, 600 kt of titania and 900 kt of liquid alum could be recovered (Aswathanarayana, 2005). In India, about 4.7 Mt/Pa of red mud is produced (6.3% of world's total) following digestion with sodium hydroxide at elevated temperature and pressure (Deelwal et al., 2014). The stripping ratio of Indian bauxite ores is higher at a ratio 1.2 as in comparison to Australia which is only 0.1 (Irshad, 2013).

2.2.5   Chromite mining

Chromite is an oxide of chromium and iron with two different chemical compositions ($FeO.Cr_2O_3$ or $FeCr_2O_4$). It contains 68% Cr_2O_3 and 32% FeO with Cr : Fe ratio of about 1.8 : 1 and occurs as a primary mineral of ultrabasic igneous rocks being normally associated with peridotite, pyroxenite, dunite and serpentinite. Chromite is important because it is the only commercial source of chromium, which is an essential element for a variety of metal, chemical and manufactured products. Many other minerals contain chromium, but none of them is found in deposits that can be economically mined to produce chromium.

About 95% of the world's chromium resources are geographically concentrated in Kazakhstan and southern Africa (Papp, 2013). The other important countries with chromite resources are Zimbabwe, Finland, India, Brazil, Turkey, Albania, Russia, the United States and Iran.

In India, chromite is mined mostly by opencast method up to a maximum depth of 63 m, except a few underground mines in Byrapur in Karnataka and Boula and Kathpal mines in Odisha. The total resources of chromite in the country (as on 1 April 2010) were estimated at 203 Mt, comprising 27% reserves (54 Mt) and 73% remaining resources (149 Mt). The leading state with chromite reserves is Orissa (93%), except some minor deposits scattered over other states like Manipur, Nagaland, Karnataka, Jharkhand, Maharashtra, Tamil Nadu and Andhra Pradesh. Gradewise, 36% resources are charge-chrome grade followed by ferro-chrome grade (19%), beneficiable grade (17%) and refractory grade (5%) and 23% accounting for low, other, unclassified and not known grades (Indian Minerals Yearbook, 2012b).

The production of chromite in India during 2011–2012 fell by 3.7 Mt representing a reduction of 13% due to falling world less demand. The number of reporting mines was 20 in 2011–2012, compared to 21 in the preceding year (Indian Minerals Yearbook, 2012b).

The management of waste dumps/overburden after chromite mining is the major environmental concern, as they alter land topography and drainage, and prevent the natural succession of plants leading to soil erosion and pollution of the environment. Chromite mining has a maximum overburden to ore ratio of 15:1.

2.2.6 Limestone and other calcareous stone mining

Limestone refers to any calcareous sedimentary rock consisting essentially of carbonates, mostly in the form of dolomite $CaMg(CO_3)_2$ or magnesite ($MgCO_3$) mixed with calcite. It is mainly used in the cement industry. It is an important raw material, often described as the world's most versatile mineral. In India, the total resources of limestone of all categories and grades (as on 1 April 2010) were estimated at 185 Gt, of which 8% (15 Gt) was under the reserve category and 92% (170 Gt) under remaining resources category (Indian Minerals Yearbook, 2012a).

In 2010–2011, limestone production was 240 Mt (CSE Report, http://www.cseindia.Org/userfiles/Mining_India.pdf), which reached at 280 Mt during 2012–2013 (IBM report, Government of India; http://ibm.gov.in/msmpmar13_07_Highlights_Eng.pdf).

Other calcareous material includes 'chalk', which is usually an extremely fine-grained, soft and friable variety of limestone.

In India, the total resources of chalk on1 April 2010 were estimated at 9.8 Mt (Indian Minerals Yearbook, 2012a).

In India, limestone mining/quarrying produces around 17.8 Mt of solid waste, with 12.2 Mt as rejects, 5.2 Mt as cuttings/trimmings (or undersize materials) and 0.4 Mt of slurry at processing and polishing facilities (http://www.tifac.org.in/index.Php?Option=comcontent SolidWasteGenerationandUtilizationinCalcareousStoneIndustry).

In general, the stripping ratio for limestone mines in India is 1 : 1.05, but the generation of overburden varies from mine to mine (Sahu and Dash, 2011). In some cases, it is as high as 1.4 te per tonne of limestone (e.g. Madras Cement Limited, Jayantipuram) to 0.54 te per tonne of limestone (e.g. ACC's unit at Jamul, Chhattisgarh) (Bhushan and Hazra, 2005).

2.3   Mining in other countries

The status of mining in major mineral-producing countries is described in this section.

2.3.1   Mining in Australia

Australia is the home of a wide range of different mineral resources and mining is a significant primary industry. The bulk of raw materials mined in Australia are exported overseas to countries such as China for processing into refined product. The major mining industries include coal and uranium, while other minerals are also extracted including iron ore, nickel, bauxite/aluminium (largest producer), gold (second largest producer after China), diamond (third largest deposits after Russia and Botswana), opal (largest producer in the world), zinc (second only to China), oil shale, petroleum and natural gas (world's third largest producer) and rare earth elements.

2.3.1.1   Coal

Australia is the world's largest exporter of coal and fourth largest producer of coal after China, the United States and India. Coal is extracted in every state of Australia but is primarily mined in Queensland, New South Wales and Victoria.

New South Wales has mostly deep mines, whereas 80% of Queensland's operations are opencast. About 70% of the coal mined in Australia is exported, mostly to eastern Asia (EIA, 2011). In the year 2000–2001, 259 Mt of coal was mined, rising to 487 Mt in the fiscal year 2008–2009 (Australia Mineral Statistics, 2009). Exports increased from 194 Mt in 2000–2001 to 414 Mt in 2011; Australia is ranked in the four biggest producers in the world (http://www.worldcoal.org/resources/coal-statistics/).

2.3.1.2 *Uranium*

Australia is the world's third largest producer of uranium after Kazakhstan and Canada with 11% of world total production. The country contains 23% of the world's proven uranium reserves but has faced opposition in recent decades due to environmental problems (Aborigines count cost of mine, 2004; Sydney International Investment Group, 2014).

2.3.2 Mining in the United States

Mining is a major industry in the United States with a production of a wide variety of including coal, copper, silver, gold and uranium are major minerals.

2.3.2.1 *Coal*

The United States holds the world's largest estimated recoverable reserves of coal and is the world's second largest producer. Coal is mined in 25 states and in 2011 production was >1 Gt (http://www.worldcoal.org/resources/coalstatistics/2013). The United States is a net exporter of coal http://www.eia.gov/coal/production/quarterly/); however, due to the displacement of natural gas by coal consumption for power production in Asia, the export of coal is increasing (Brown, 2013).

2.3.2.2 *Copper*

Copper mining has been a major industry since the rise of the northern Michigan copper district in the 1840s. The United States produced 1.2 Mt of copper in 2012, worth $9 Bn, and is the world's fourth largest producer after Chile, China and Peru (Papp 2013).

2.3.2.3 *Silver*

Though the silver industry declined greatly after the demonetization of silver in the late 1800s (by the Coinage Act of 1873 –and known pejoratively as the 'Crime of 73'), there has been an increased in production in the last decade with 1.2 kt being produced in 2007.

2.3.2.4 *Gold*

The United States is the third largest gold-producing nation, behind China and Australia. Most of the gold produced today comes from large open-pit heap leach mines in the state of Nevada. In 2012 production was 230 t (United States Geological Survey (USGS), 2013).

2.3.2.5 *Uranium*

The United States was the world's leading producer of uranium between 1953 until 1980, when annual production peaked at 16,810 t U3O8. Until the early 1980s, there were active uranium mines in Arizona, Colorado,

New Mexico, South Dakota, Texas, Utah, Washington and Wyoming (Finch et al., 1973). In 2012, the United States produced 4.1 million pounds (1860 t) of uranium oxide concentrate (US Energy Information Administration, 2013), primarily for use in nuclear power generation.

2.3.3   Mining in South Africa

Mining in South Africa has been the main driving force behind development of Africa's most advanced and richest economy. South Africa is a home of major and minor mineral reserves, including diamond, gold, chrome, manganese, platinum, vanadium, vermiculite, ilmenite, palladium, rutile, zirconium, iron and coal. It is the world's largest producer (Mineral Commodity Summaries) of chrome, manganese, platinum, vanadium and vermiculite and second largest producer of ilmenite, palladium, rutile and zirconium.

2.3.3.1   *Coal*

South Africa is ranked as the seventh largest producer of coal in the world, but the third largest exporter. In 2011 it is estimated that South Africa produced more than 255 Mt of coal (BP, 2012) and consumed 75% of this domestically (Production and consumption of coal-2003 estimates). Around 77% of South Africa's energy needs are derived directly from coal (Department of Energy, South Africa, 2012) and 92% of coal consumed by the African continent is produced in South Africa (International Energy Outlook, 2009).

2.3.3.2   *Iron ore*

In 2012, South Africa overtook India to become the world's third biggest iron ore supplier to China, the world's largest consumer of iron ore (International: Mining Weekly, 2013).

2.3.3.3   *Gold*

South Africa holds about 50% of the world's gold reserves. However, production has declined and since 1993, falling behind the China, Australia, the United States, Russia and Peru. In 2011, the production of gold was 7% of the world's total (http://www.iser.uaa.alaska.edu/Publications/mining-indicators.pdf) falling to 6.1% in 2012 (USGS, 2013). Two mines are the deepest mines in the world (the East Rand Mine, Boksburg and TauTona in Carletonville), extending to depths of 3585 m, and, 3900 m, respectively.

2.3.3.4   *Diamond*

Although the percentage of the world's diamonds produced by South Africa has fallen to 12%, the country is still the world leader, with reserves located in seven large mines around the country (controlled by the

De Beers Consolidated Mines Company). In 2003, De Beers operations accounted for 94% of the nation's total diamond output of 11.9 million carats, rising to 15.8 million carats (3.2 t) in 2005, and then falling to 7 million carats in 2011 (Chamber of Mines of South Africa, 2012).

2.3.3.5 Platinum and palladium

South Africa is the largest producer of platinum and similar metals (CIA World Fact book). In 2005, 78% of the world's platinum was produced in addition to 39% of the world's palladium. The South African platinum industry is facing a number of challenges (Jansen, 2012) leading to a fall in production by 2011 (source: SFA, Oxford, Johnson Matthey, KPMG analysis; Jansen, 2012). In 2005, over 163 t (5,200,000 oz) of platinum was produced generating export revenues of $3.8 billion USD (USGS, 2005).

2.3.3.6 Chromium

The country holds about 70% of the world's total chrome reserves, mostly located in the Bushveld Igneous Complex (BIC) ores, and produces 75% of the world's ferrochrome. Chromium is mined at ten for use in stainless steel and other industrial applications. South Africa's produced 100% of the world's total in 2005, amounting to 7.5 Mt (USGS, 2005). In the year 2009, an estimated 9.6 Mt of chromium ore was produced (41% of world production), declining by 2% in 2010 (Papp, 2011).

2.3.4 Mining in the United Kingdom

The United Kingdom has a rich history of mining with evidence of mining in Wales dating back to the Bronze Age (2200–850 BC) (Pearce and Armfield, 1998). Mining in the United Kingdom produces a wide variety of fossil fuels, metals and industrial minerals, which has significantly contributed to the country's economy.

2.3.4.1 Coal mining

Coal was the main source of UK energy until the late 1960s, with highest production being 228 Mt in 1952. Ninety-five per cent of this came from around 1334 deep mines that were operational at that time, with the remainder from 92 surface mines (Coal Statistics, 2013). However, with more diversification of UK energy market the production of coal significantly declined, and in 1983 the UK stopped being an exporter of coal.

In 2013, surface coal production in the United Kingdom fell to 13 Mt; being the lowest ever recorded. In 2011 and 2012, due to rising gas prices coal use increased, but in 2013, total demand fell to 60 Mt (Coal Statistics, 2013), with around 71% supplied via imports, with the remaining 29% met by the indigenous sources (2009 figures).

In the United Kingdom, coal is primarily extracted by opencast (surface) mining methods from seams up to 5 m in depth. Additionally, a small amount of coal is recovered from tip washing (British Geological Survey Commissioned Report, 2010). Surface mining involves a quarrying down to 100 m (exceptionally to 200 m), with large amounts of overburden being generated. The stripping ratios are variable, but currently, ratios of 20:1 are economic (British Geological Survey Commissioned Report, 2010).

2.3.4.2   Iron ore copper, tin, lead and silver mining

Little copper or iron are mined today (British Geological Survey Commissioned Report, 2010). However, some tin and lead are produced from deep mines in England, Wales and Scotland.

2.3.4.3   Industrial minerals

The United Kingdom is well endowed with a wide range of industrial minerals which are used in the manufacturing of chemicals, ceramics, plaster products, glass. Minerals such as kaolin, ball clay and potash contribute significantly to export markets (UK Minerals Forum Working Group 2013-14-Future Mineral Scenarios for the UK, 2014).

2.3.4.3.1   Kaolin

The United Kingdom is the world's third largest producer of kaolin after the United States and Brazil. Kaolin is a fine white-coloured clay formerly known as China clay. Kaolin is a hydrated aluminosilicate and is different to other kaolinitic clays (ball clay or fire clay) due to its whiteness, its particle size/shape and its rendering opacity.

Kaolin has a range of industrial applications including the production of paper (50%), ceramics (30%) and other speciality applications (20%), including in paint, rubber, plastics, adhesives/ sealants, pharmaceuticals, animal feed, white cement and in glass fibre (UK Minerals Forum Working Group 2013-14-Future Mineral Scenarios for the UK, 2014).

In Britain, the kaolin resources are confined to the granites of the south-west England. For commercial reasons, a figure for existing kaolin reserves is not available. However, sufficient proved reserves of kaolin exist in Cornwall and Devon pits to sustain current rates of production for at least 50 years.

Each tonne of marketable kaolin produces up to 9 te of other materials (4 te of sand, 3 te of rock/stent, 1 te of overburden and 1 te of micaceous residues).

2.3.4.3.2   Ball clay

Ball clays are highly plastic and fire to a near white colour. They primarily consist of three minerals – kaolinite, mica and quartz – and are used in the manufacture of whiteware ceramics (sanitaryware, wall/floor tiles and tableware). The United Kingdom is a leading world producer and exporter of high-quality ball clay, being produced in three small basins in Devon and Dorset. Worldwide, British ball clay exports contribute to over 50% of global sanitaryware production (British Geological Survey Commissioned Report, 2011). As the by-products, ball clay extraction produces interburden and overburden sand and small amounts of lignite.

2.3.4.3.3 Carbonate rocks (limestone, chalk and dolomite)

The United Kingdom produces carbonate rocks in abundance – limestone (principally calcium carbonate ($CaCO_3$)), chalk (very fine-grained limestone) and dolomite (calcium magnesium carbonate ($CaMg(CO_3)_2$)). The primary application of limestone in the United Kingdom is in the construction industry and also as an essential raw material for cement manufacture and building stone. In addition, industrial limestone is an important raw material in iron, steel, glass manufacturing, sugar refining and soda ash manufacturing and is also used as a mineral filler in papers, paints, plastics, rubbers, pharmaceuticals and cosmetics. However, in the United Kingdom, one very important area of utilization of limestone is for environmental applications via water and effluent treatment, flue gas desulphurization and pollution control.

In the United Kingdom, the total demand for limestone and chalk for industrial and agricultural use in 2004 was 6.9 Mt and 2.1 Mt, respectively. In British market, quicklime (CaO – about 1.7 te of limestone is burnt to produce 1 te of quicklime) is mostly sold and the production of lime is around 2.5 Mt year^{-1} (British Geological Survey Commissioned Report, 2006). Tunstead Quarry in Derbyshire is the largest producer of lime and chemical stone, while Melton Ross in North Lincolnshire produces highest chalk in the United Kingdom.

2.3.5 Mining in China

China is having rich reserves of various minerals and is the leading producer of several major minerals in the world. The major mining in China is done for coal and gold.

2.3.5.1 *Gold*

The People's Republic of China emerged as the world's largest gold producer with extensive gold mining. For the year 2007, gold output rose 12% from 2006 to 276 te (or 9.7 million ounces) to become the world's largest for the first time. Because of declined production of gold in South Africa by 50% in the past decade, China overtook South Africa, which was the largest producer for 101 years since 1905. In recent years, China's gold mining industry has received increased foreign and domestic investment because of production of nearly 300 te of gold in 2008. China has been the world's largest producer from 2006 to 2012 with 403 te of gold (produced in 2012) representing an increase of 11.3% over the previous years' total (USGS, 2013).

2.3.5.2 Coal

China is the largest coal producer and consumer in the world, with major reserves lying in northern China in Shanxi Province. In 2011, China alone accounted for 49% of the global demand for coal (BP p.l.c., 2012). Coal from southern mines tends to be higher in sulphur and ash and, therefore, unsuitable for many applications (Energy Information Administration, 2006).

Coal is the major source of energy in China and its demand continues to increase. On 6 July 2008 in central and northern China, 2.5% of the nation's coal plants (58 units or 14 MW of capacity) had to shut down due to coal shortages. This forced local governments to limit electricity consumption and issue blackout warnings. The shortage is somewhat attributed to the closing of small dangerous coal mines (Ying, 2008). In 2011, China accounted for 48.2% of the global demand for coal. Despite China's twelfth 5-year plan to reduce its reliance on coal, it remains the backbone of the country's energy generation (Censere Report, 2012). According to global coal statistics of 2011, China produced 3471 Mt of coal and ranked number one in the world (http://www.worldcoal.org/resources/coal-statistics/2013).

A list of top five coal and brown coal-producing countries is given in Tables 2.2 and 2.3.

Table 2.2 Top five coal producers (2011).

PR China	3471 Mt	Russia	334 Mt
United States	1004 Mt	South Africa	253 Mt
India	585 Mt	Germany	189 Mt
Australia	414 Mt	Poland	139 Mt
Indonesia	376 Mt	Kazakhstan	117 Mt

Source: http://www.worldcoal.org/resources/coal-statistics/, accessed 13 August 2013.

Table 2.3 Top five brown coal producers (2011).

Germany	176 Mt	Australia	69 Mt
China	136 Mt	Poland	63 Mt
Russia	78 Mt	Greece	59 Mt
Turkey	74 Mt	Czech Republic	43 Mt
United States	74 Mt	India	41a Mt

Source: http://www.worldcoal.org/resources/coal-statistics/, accessed 13 August 2013.

2.4   Types of mine waste disposal

In general, there are four ways in which mine waste is produced:

(1) Overburden: the soil and rock that is removed to gain access to a mineral resource.
(2) Waste rock: rock that does not contain enough mineral of economic interest.
(3) Tailings: these are finely ground host rocks from which the desired minerals have largely been extracted. The residue takes the form of slurry of ground-up ore that remains after minerals have been largely extracted.
(4) Heap leach spent ore: here, the crushed ore is placed on a membrane-lined pad and irrigated with the appropriate reagent, including sulphuric acid, in the case of copper or uranium; the rock remaining in the heap leach after the ore has been extracted (http://india.indymedia.org/en/2002/12/2456.shtml).

Mine wastes are generated during the process of extraction, benefi-ciation and processing of minerals. Extraction is the first phase that consists of the initial removal of ore from the ground. This is generally achieved by blasting, producing huge volume of waste (soil, debris and other materials). This debris is often unwanted and is normally stored in stockpiles within the mine and, occasionally, on public land. The big-ger the mine, the greater is the quantum of waste generated. Opencast mines are, therefore, more pollution intensive as they generate much higher quantities of waste compared to the underground mines. Open-pit mines produce 8–10 times as much waste as underground mines (Anon, 2006).

Waste rock is an unused extraction product containing metal content, which is too low and difficult to recover economically. It is usually stored in landfill sites near the mine site, so as to save the transportation cost (Sartz, 2010). Usually, opencast pits and quarries cause more mine waste generation than an underground mine. Waste rock mainly exposes shallow ore after surface stripping. Huge amounts of waste rock are generated everyday by mining and quarrying activities (Sartz, 2010). Disposal of mine waste from hard rock metal mines is the largest envi-ronmental problem.

Tailings are processed waste, usually fine ground rock, from which valuable materials have been extracted. Depending on the ore properties, technologies used are very different and different units (leaching, floatation) can produce different kinds of wastes. Tailings consist of the materials formed after enrichment process of the ore and are normally stored in impoundments. Mine tailings contain high concentrations of toxic and carcinogenic metals, like Pb, Cu, As, Cd, Cr, Zn, Ni and Hg (Nehdi and Tariq, 2007; Yang et al., 2009). These metals are nondegradable and are persistent and may pollute surface and subsurface water resources and, thus, pose a serious problem to public health (Nehdi and Tariq, 2007; Yang et al., 2009; Sartz, 2010).

Different physical and chemical processes take place in mine waste rock and tailings (Sartz, 2010). Oxygen is diffused through the waste rock piles due to the difference in oxygen concentrations between the inner layer of piles and surrounding air and this transformation of gas occurs by convection and diffusion mechanisms (Sartz, 2010). Geochemical conditions can also affect the oxidation process (Nehdi and Tariq, 2007).

The large volume of tailing generated creates challenging management issues. The quantities of tailings produced at most precious and base metal mines are about the same as the quantity of that is mined. For example, a mine producing 100 kte of copper ore per day also produces close to 100 kte of tailings. This equates to several thousand truckloads per day of tailings that needs to be managed and disposed of in an environmentally responsible manner.

Most adverse environmental impacts of tailings are related to their impact on water resources. As tailings are consolidated, a supernatant separates from the slurry, forming a surface pond, from where it can be decanted for disposal or, more commonly, for reuse. In tropical regions, where precipitation exceeds evaporation, decanted water may carry significant amount of tailing fines. Once released to the environment, these fines may be carried over long distances before eventually settling as soft sediments. Sometimes associated with fine particles are elevated concentrations of trace metals that can be mobilized by natural chemical or biological processes.

Even when all tailing solids are contained, discharged water may remain harmful to the environment. Even the infiltering water emanating from and passing through tailing storages may become acidic with elevated concentrations of dissolved trace metals, which may impact the wildlife or livestock through drinking water from tailing ponds.

Considering all the main minerals produced in the country, around 1.9 Gt tonnes of overburden and waste materials was generated to excavate only 750 Mt of minerals (Table 2.2) in only 1 year, that is, 2005–2006. In 7 years between 1999–2000 and 2005–2006, the total quantum of waste generated was 10.8 Gt of waste. Among all the key minerals, coal mining generated maximum overburden, and as expected, the amount of overburden generated has increased over the years along with the increase in mineral extraction.

2.5 Wastelands

Wastelands are the areas of unproductive land unable to yield even half of its potential. Degraded or wastelands have been defined by various ministries and the Planning Commission in different ways. The Committee constituted by the Planning Commission in 1987 to deal with the definition of wastelands defined wastelands as 'Degraded land which can be brought under vegetative cover with reasonable effort and which is currently under-utilized and land which is deteriorating for lack of appropriate water and soil management or on account of natural causes. Wastelands can result from inherent/imposed disabilities, such as by location, environment, chemical and physical properties of soil or financial or management constraints.'

At present, approximately 68 million hectares of land in India is considered wasteland. Of these lands, approximately 50% are non-forest lands, which can be made fertile again if treated properly (*State of Environment Report*, 2009). The removal of forest can result from fire, utilization for fuel and by the dumping of industrial hazardous wastes, including by tailings.

The Indian scenario of environment and forests continues to cause concern. Destruction and degradation of forests are taking a heavy toll of soil and water resources. An estimated 6 Gt of topsoil with essential nutrients is flowing into the sea every year. Loss of topsoil, vegetative cover and unregulated surface run-off seriously affect the society. Overall degradation of nature is also making our resources less productive, leading to impoverishment of the rural population (eighth 5-year plan, Vol-2).

In order to ensure food security, environmental protection and biodiversity conservation for ever-increasing population, wasteland needs to be brought under cultivation, the productivity of existing agricultural land needs to be enhanced, and suitable measures need to be taken up to prevent the fertile cultivated land from degradation. Wastelands which are currently lying either unutilized or partially utilized may contribute significantly to food security (http://pib.nic.in/release/release.asp?relid=63602).

In recent years, an unprecedented increase of mine waste dumps in India has been of great concern, not least for their impact on ecosystems. Poor handling practice generates overburden dumps, which are extremely fragile and difficult to revegetate. Mine spoils offer adverse conditions for soil microbial and plant growth, due to their low organic matter, unfavourable chemistry, poor structure (coarseness/compaction) and their isolation from vegetation. These conditions are generally unfavourable for ectomycorrhizal fungal growth which plays an important role in the regulation of plant communities, net primary productivity and nutrient cycling. Poor microbial populations inhibit plant growth and the process of ecological succession, leading to delayed recovery of damaged land.

Some disposal methods, such as conical dumping by dragline, have resulted conditions that are susceptible to gully erosion and dump slope failure. This results in air and water pollution, reduced aesthetic values, blockage of pit access and filling of sump and thereby problem of pumping leading towards flooding of the working area during rainy seasons. In rainy season, soil often creeps down due to sandy loam texture of certain overburden dump materials and prevents the invasion and establishment of early and late successional plant species. Physical disturbance causes retrogression of the system into earlier succession stages and creates habitats for disturbance-tolerant and fugitive species, whereas disturbance-sensitive species are eliminated (Rapport et al., 1985). Therefore, development of vegetative cover over dump slopes makes the slope stabilized against soil erosion. A scientific, rational and sustainable utilization of natural resources and site protection from toxic contaminant releases is vital for sustainable industrial growth and socioeconomic development of the country. It helps the proper functioning of the ecosystems, and this measure needs to be considered during the mine planning stage.

2.6   Waste generation

Population growth, increasing urbanization and rising living standards and technological innovation have contributed to an increase both in the quantity and variety of solid wastes generated by industrial, agricultural and mining activities. Despite the expansion of the global waste-to-energy (WTE) industry in the past decade, hundreds of millions of tonnes of municipal solid wastes (MSW) still ends up in landfills. Worldwide mine activities produce a considerable amount of sulphidic tailings, and this phenomenon can expose the buried ore to the environment (Méndez-Ortiz et al., 2007). It is estimated that for every tonne of waste that is landfilled, greenhouse gas emissions in the form of carbon dioxide increased by at least 1.3 te (http://www.waste-management-world.com/articles/2003/07/an-overview-of-the-global-waste-to-energy-industry.html). Production of coal and other minerals and generation of overburden in Indian mining areas from 1999–2001 to 2005–2006 are given in Table 2.4.

In India, about 960 Mt of solid waste is being generated annually as by-products during industrial, mining, municipal, agricultural and other processes. Since India produces 89 different minerals, these

Table 2.4 Coal and other mineral production and overburden generation (Mt) in Indian mining areas from 1999–2001 to 2005–2006.

	1999–2001	2000–2001	2001–2002	2002–2003	2003–2004	2004–2005	2005–2006
Coal							
Production	300	310	323	337	356	377	407
Overburden	1100	1135	1183	1235	1304	1383	1493
Bauxite							
Production	7.1	8.0	8.6	9.9	10.2	12.0	12.3
Overburden	4.3	4.8	5.20	6.0	6.6	7.2	7.5
Limestone							
Production	129	123.6	129.3	155.7	153.4	165.8	170.4
Overburden	135	129.4	135.3	163.0	160.5	173.5	178.3
Iron ore							
Production	75.0	80.7	86.2	99.1	122.8	145.9	154.4
Overburden	69.9	75.2	80.3	92.3	114.5	136.0	143.9
Others							
Production	7.5	5.0	4.9	7.8	8.3	9.4	9.4
Overburden	14.5	10.6	10.4	15.2	16.3	19.3	18.6

Source: http://www.cseindia.org/programme/industry/mining/political_minerals_mapdescription.htm.

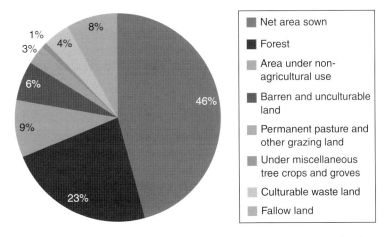

Figure 2.1 Land-use classification in India (2011–2012). (Source: *Agricultural Statistics at a Glance* (2013).)

result in diverse and potentially hazardous wastes, with potential for considerable environmental harm. Of these, approximately 350 Mt is organic wastes from agricultural sources, 90 Mt is inorganic waste of industrial and mining sectors, and 4.5 Mt is hazardous in nature (Pappu et al., 2007). Nevertheless, an estimated 147 Mha area suffers from various forms of land degradation due to water and wind erosion and other complex problems like alkalinity/salinity and soil acidity due to waterlogging (Figure 2.1) (*State of Environment Report*, 2009).

Table 2.5 The statewise burden of mining wastes in India.

State	No. covered	Total geographical area of districts covered (Mha)	Total area in covered districts (Mha)	Percentage of total geographical area
Andhra Pradesh	23	27.51	5.17	18.81
Arunachal Pradesh	13	8.37	1.83	21.88
Assam	23	7.84	2.0	25.52
Bihar	55	17.38	2.1	12.08
Goa	2	0.3702	0.61	16.57
Gujarat	25	19.6	4.3	21.95
Haryana	19	4.42	0.3733	8.45
Himachal Pradesh	12	5.56	3.16	56.87
Jammu and Kashmir	14	10.13	6.54	64.55
Karnataka	27	19.17	2.08	10.87
Kerala	14	3.88	0.14	3.73
Madhya Pradesh	62	44.34	6.97	15.72
Maharashtra	32	30.76	5.34	17.38
Manipur	9	2.23	1.29	58.00
Meghalaya	7	2.24	0.99	44.16
Mizoram	3	2.11	0.41	19.31
Nagaland	7	1.66	0.84	50.69
Orissa	30	15.57	2.13	13.71
Punjab	17	5.04	0.22	4.42
Rajasthan	32	34.22	10.56	30.87
Sikkim	4	0.71	0.36	50.30
Tripura	4	1.05	0.13	12.17
Tamil Nadu	29	1.30	2.30	17.70
Uttar Pradesh	83	29.44	38.77	13.17
West Bengal	18	8.87	0.57	6.44
Union Territory	20	1.09	0.057	5.23
Total	584	31.66	63.85	20.17

Source: 1 : 50 000 scale wasteland maps prepared from Landsat Thematic Mapper/IRS LISS II/HI Data. http://dolr.nic.in/WastelandsAtlas2005/Wasteland_Atlas_2005.pdf, accessed 27 June 2015.

Table 2.6 Waste generation trends in India.

Year	Per capita waste generation (g day^{-1})	Total urban municipal waste generation (Mt year^{-1})
1971	375	14.9
1981	430	25.1
1991	460	43.5
1997	490	48.5
2025	700	Double the amount of 1997

Source: www.indiaenergyportal.org/energy_stats.php.

Globally, the estimated quantity of total waste generation was 12 Gt in the year 2002, of which 11 Gt was industrial wastes and 1.6 Gt was MSW. About 19 Gt of solid wastes is predicted pa by 2025 (Yoshizawa et al., 2004). Annually, Asia alone generates 4.4 Gt of solid wastes and MSW comprise 790 Mt, of which about 48 (6%) Mt is generated in India (Central Pollution Control Board (CPCB), 2000; Yoshizawa et al., 2004). By the year 2047, MSW generation in India is expected to reach 300 Mt and land requirement for disposal of this waste would be 170 km^2. This compares with 20.2 km^2 that was used in 1997 for the management of 48 Mt of waste (CPCB, 2000). In 2006, 1.8 Gt of waste was generated from mining of major minerals. Tables 2.5 and 2.6, respectively, show the burden of wastes and waste generation trends in India.

Mining activities in China have generated a total derelict land of about 3.0 Mha by the end of last century (Wong and Luo, 2003), and the figure is increasing at an alarming rate of 46 700 ha year^{-1} (Bai et al., 1999). Some 280 000 mines were in operation in 1994, and 5 million people were employed in the mining sector (UNEP, 1997). Of these mines, over 8000 were state-owned mining enterprises and around 230 000 were small-scale mining workshops (Shu et al., 2003) throughout the country, many of which were located in the poor and remote areas. Mining operations and their associated mine tailings polluted a land area of 600 000 ha (Li, 2006), causing a direct economic loss of over 9 billion yuan and an indirect loss of about 30 billion yuan pa (Liu and Shu, 2003).

As far as waste generation from coal mining is concerned, 3–4 te of waste is generated for every tonne of coal extracted. In the last decade, coal mining in China has degraded the quality of land of an estimated 3.2 million hectares (Greenpeace International Report, 2010). The overall restoration rate (the ratio of reclaimed land area to the total degraded land area) of mine wasteland was only about 10–12%. Similarly, in the United States, between 1930 and 2000, coal mining altered about 2.4 million hectares (5.9 million acres) of natural landscape, of which most were originally the forests.

2.7 Solid waste generation

In India, more than 200 Mt of non-hazardous inorganic solid wastes is being generated every year (CPCB, 2000; Saxena and Asokan, 2002), out of which 80 Mt is mine tailings/ores of iron, copper and zinc mines (Gupta, 1998; Agrawal et al., 2004). The mine wastes dumped over the land surface can be seen in Figure 2.2. The country has considerably economically useful minerals that constitute one quarter of the world's known mineral resources (Table 2.7).

(a)

(b)

Figure 2.2 Mine overburden dumps. (a) Iron ore and (b) coal.

Table 2.7 Mining wastes in India (1999–2006) (in Mt).

	1999–2000	2000–2001	2001–2002	2002–2003	2003–2004	2004–2005	2005–2006
Coal							
Production	300	310	323	337	356	377	407
Overburden and other wastes	1100	1135	1183	1235	1304	1383	1493
Bauxite							
Production	7.1	7.99	8.59	9.87	10.92	11.96	12.34
Overburden and other wastes	4.3	4.84	5.20	6.0	6.6	7.2	7.5
Limestone							
Production	129	123.6	129.3	155.74	153.39	165.75	170.38
Overburden and other wastes	135	129.4	135.3	163	160.5	173.5	178.3
Iron ore							
Production	75.0	80.7	86.2	99.1	122.8	145.9	154.4
Overburden and other wastes	69.9	75.2	80.3	92.3	114.5	136	143.9
Others							
Production	7.5	5.0	4.9	7.8	8.3	9.2	9.4
Overburden and other wastes	14.5	10.6	10.4	15.2	16.3	19.3	18.6
Total overburden and other wastes from major minerals (excluding beneficiation and processing wastes)	1324	1355	1414	1512	1602	1719	1841

Source: *State of India's Environment – A Citizens Sixth Report* (2008).

2.8 Ecological degradation and disturbance

Ecological degradation has a long history but was considered to be a problem only by a small group of nature conservationists, scientists and professional ecologists. The awareness of wider public, and the beginning of the general environmental movement, is usually traced back to the publication of Rachel Carson's Silent Spring in 1962. This global awareness was further enhanced, along with firm commitment by national government, by the conference on biodiversity in Rio de Janeiro in 1992. Since then, biodiversity has been recognized as a political goal, and many countries have signed treaties and implemented programmes to prevent further losses. Social activities that are detrimental to biodiversity and to environmental health have been identified, and policies are being developed to halt such activities or counteract their consequences.

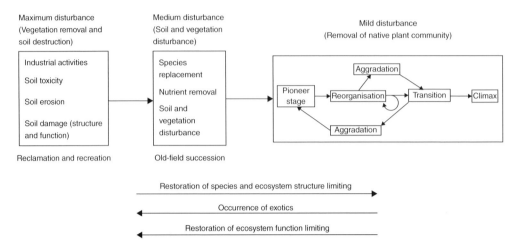

Figure 2.3 Categorization of degradation and restoration (Jordan et al., 1990).

Forest harvesting is one of the most severe types of degradation from mining and industrial operations which particularly drawn the attention, because its effects are so drastic, in and around the mining areas. It has been a particular challenge, because the original ecosystems have been totally destroyed. In severely disturbed ecosystem, such as those resulting from surface coal mining, geology–soil–plant stability circuit is disrupted; flora, fauna, hydrological relations and soil biological systems are drastically disturbed, and if left uncared, different types of degradation accumulate, with drastic consequences. So the need for restoration as an integral part of the philosophy and activities of all human societies is crucial (Cairns, 1995).

Degradation can be categorized in different stages on the basis of the increasing levels of disturbance. The first and mildest level of disturbance involves disruption or removal of native plant community, without severe disturbance of the soil. This is labelled 'secondary succession' (Jordan et al., 1990) seen in Figure 2.3.

The different boxes within this stage represent different successional stages within a community type. The arrows indicate several possible paths of succession, indicating that 'succession' is not a one-directional process and that it can be deflected significantly by disturbance or even chance. The distinctive feature in this stage of degradation is that the soil is left largely intact. Thus, growing conditions may be close to those in undisturbed soil, and it may be possible to directly establish native species selected from mature 'target communities'.

In its earliest stages, the restoration of a profoundly disturbed site (left) involves restoration of function and tends to be ecosystem oriented. At later stages, restoration is community oriented and involves species composition, population dynamics and species interactions. To some extent, this scheme may provide a way of identifying processes of interest.

The second stage of site degradation includes information on species–site interactions, when both the vegetation and the soil have been damaged. An example might be an abandoned (old field) site, subjected to ploughing and other agricultural purposes for some time. Here, removal of nutrients in crops along with possible erosion and compaction may have lowered both nutrient and water availability. The process of 'old field succession' by which a sequence of species replacements rebuilds both vegetation and soil is one of the classic areas of ecological research.

The third category of disturbance includes those areas where the vegetation is completely removed and the soil is converted to a form totally outside the range of natural conditions. This is the most severe case where the measurement of environmental conditions and species tolerances and requirements becomes even more crucial. In this case, the soil needs to be recreated before the mature plant community may function.

The soil alteration itself can be accomplished in one of two ways: the first is by physical, chemical or mechanical site preparation and another way is to use plants. The introduction of exotics into restoration processes should be restricted, because their removal may pose major problems in the final control of species composition.

An ecosystem has two major attributes – structure and function – each made up of different elements. They can be used to define and illustrate the damage that ecosystems can suffer (Magnuson et al., 1980; Bradshaw, 1987b). An original ecosystem will typically (although not always) have high values for both. Degradation drives one or both attributes downwards, often to nearly nothing. If the area is left to its own devices, the natural processes of primary succession will restore the ecosystem to its starting point (Miles and Walton, 1993).

Disturbances are referred to as 'severe' if they result in the complete loss of native soil and vegetation and if they disrupt or destroy natural surface and subsurface hydrological pathways. When wild land areas become severely disturbed, especially by human activities, various natural ecosystem processes are often destroyed or greatly altered, thus leading to degradation in natural resource quality and quantity. Many severe human-caused disturbances lead to major alterations to ecosystems by greatly simplifying their structure and function (Rapport and Whitford, 1999). Disturbed areas, left to recover naturally, often degrade further because natural recovery processes cannot resume, due to soil weathering and erosion. At least three consequences exist to not directly implementing restoration following human disturbances, as suggested by Rapport and Whitford (1999):

(1)　Disturbed systems become vulnerable to invasive, often exotic and highly competitive species with alternative adaptations that block or hinder the re-establishment of native species.

(2)   Disturbance of natural soils and the loss of their complex structural and chemical properties limit the ability of the system to support native organisms, thus leading to greater instability.
(3)   Disruption of nutrient cycling changes the character of the system and disperses the nutrient base capable of supporting native organisms.

Often, the net effect of continued degradation is a loss of productivity, diminished biodiversity and loss of resilience in the system to recover (Brown and Amacher, 1999; Rapport and Whitford, 1999).

Some disturbances are so severe that natural recovery processes can become deflected, suspended or terminated altogether. Under these conditions, natural recovery is extremely slow or does not occur at all. Areas often left to recover by natural processes include abandoned mines on public lands. In many cases, waste rock and debris containing toxic chemicals have been exposed to surficial weathering, resulting in residual material bearing little resemblance to natural soil. Mining and mineral processing often concentrates such materials in the form of spoils and tailings. Erosion from these spoils, especially where pyritic materials and other chemicals are found, results in sediment transport leading to acid, metal or other chemical contamination of adjacent downslope plant communities, waters and riparian and aquatic ecosystems (Amacher et al., 1993; Johnston et al., 1975).

The physical and chemical properties of toxic spoil material frequently exceed the physiological tolerance of virtually all vascular plants and thus inhibit or completely retard the establishment of plant seedlings and other processes of natural succession. Consequently, erosion and other degradation pressures continue, sometimes for decades, leading to degradation of water quality and other resources.

Restoring native wild land communities that have been severely disturbed is a critical challenge for land managers (Brown and Chambers, 1990) to reinitiate natural succession and other recovery processes to reverse effects of severe disturbances and to repair natural resource integrity. Through restoration, the self-sustaining ecological processes responsible for ecosystem stability, diversity, productivity and resilience that were disrupted or destroyed are re-established (e.g. MacMahon and Jordan, 1994; Cairns, 1995).

Succession can be considered the natural universal process of ecosystem genesis and development responsible for the evolution, formation and development of ecosystem components such as flora, fauna, microorganisms, soils, nutrient cycling, hydrological properties as well as other constituents and their interactions. Natural succession is reinitiated on disturbed land to serve as the primary 'driver' for developing and re-establishing natural processes, which include both the abiotic and biotic components, required to sustain the system consistent with current climatic and other environmental conditions.

Coal-mining activities result into two types of land disturbances: derelict land covered with the waste from underground mining and opencast mined land which is excavated and later replaced. When the mines are closed, the areas near mines are converted into 'derelict land', defined as 'land so damaged by industrial or other developmental activities that it is incapable of beneficial use without treatment'. The broad definition of derelict land could be applied to 'contaminated land', although this term is used specifically for land affected by chemical substances. The treatment of such land to bring it back into use is different in two mining methods. Derelict land in general undergoes 'reclamation', whereas 'remediation' is carried out on contaminated land. In contrast to the creation of derelict land by the deep mining of coal, surface or opencast mining only leads to temporary land disturbance.

2.9 Restoration ecology and ecological restoration

Restoration involves the construction of stable ecosystems for a long term, so it is necessary to ensure that current production does not come at the expense of future nutrient availability. The practice of ecological restoration and the science of restoration ecology have emerged as the major tools available to humankind for mitigating, arresting and reversing the adverse effects of human activity on the earth system, particularly since the industrial revolution.

The challenges that we face include:

- Food, water and energy security
- Loss of biodiversity
- Global climate change
- Rising sea level

The first two are a result of direct human pressures, with increasing population sizes demanding more resources; the last two are the result of amplification of the positive loops of natural feedback mechanisms, with the production of greenhouse gases from fossil stocks as the forcing function (Harris and Diggelen, 2006).

Forests are considered to be resilient, and if given sufficient time and duration by the cessation of disturbances, degraded land will revert to forest. Forest ecosystem undergoes dynamic processes, subject to natural development as well as natural and biotic disturbances.

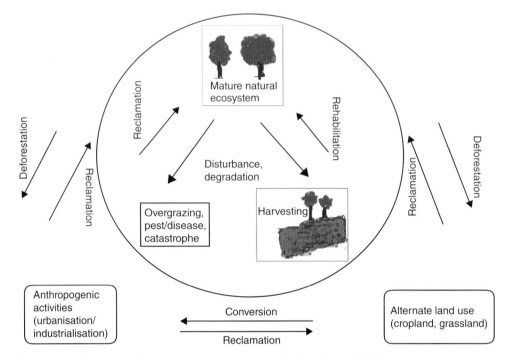

Figure 2.4 The process of ecosystem restoration (Stanturf and Madsen, 2002).

The simplest conceptualization of the relationship between degradation and recovery processes is to place a forest from a continuum from
a natural to degraded state (Bradshaw, 1997; Harrington, 1999).
Ecological factors including biomass or biodiversity in a forest subjected to degradation follow a linear trajectory. At any point along the
trajectory, recovery towards a natural forest can be initiated once the
stress or disturbance abates. Recovery of disturbed ecosystem to a more
natural forest will take long duration; however, human intervention
can accelerate the reversion process. This has been observed by Tripathi
and Singh (2008), who examined revegetated mine spoils along an age
gradient that natural recovery takes much longer time in restoration as
compared to re-vegetation.

Restoration is any end point within the natural range of managed
forests, where self-renewal processes operate (see Figures 2.3 and 2.4)
beginning with degraded forests (rehabilitation) or after deforestation
or conversion to alternate land uses (reconstruction or reclamation).
Self-renewal processes operate within forests that are disturbed but
not degraded (regeneration/reforestation) (Figure 2.4) (Stanturf and
Madsen, 2002).

A conceptual framework for forest restoration has a starting point of
a degraded forest (A) and an idealized end point of a forest restored to
natural (prior to disturbance) end point (*). The symmetric degradation/
recreation trajectories have intermediate points that represent staring/

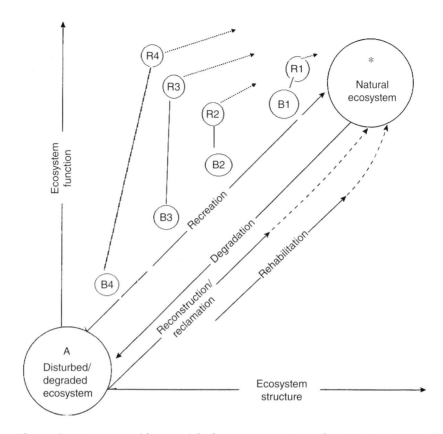

Figure 2.5 A conceptual framework for forest restoration (Stanturf and Madsen, 2002).

ending points (B₁–B₄) for reconstruction or reclamation of severely degraded forests (deforested and converted to other land use) or less severely degraded forests (rehabilitation). Recovery/replacement trajectories (R₁–R₄) denote restored ecosystems, which are close to the natural system but lack the original structure (species composition) and function of native forests (Figure 2.5).

For both reconstruction and rehabilitation, continuing intervention over time may move to forest condition closer to the natural end point (Figure 2.5). Rehabilitation encompasses many techniques to restore stand structure, species composition and natural disturbance regimes or to remove exotic plants. Specific forms of rehabilitation are termed conversion (Nyland, 2003; Spiecker et al., 2004) or transformation (Kenk and Guehne, 2001). Rehabilitation of degraded forests has one of the intermediate conditions (B₁–B₄) as a starting point; forest cover has been removed or degraded, but no change to non-forest land use has occurred.

Management of disturbed land to hasten its regeneration through reclamation is merely the management of ecological processes for a specific purpose/goal. Ecological restoration is a tool to improve the quality of degraded ecosystems and is now carried out on an ever-increasing scale. In general, restoration needs to be considered in terms of the interaction between species and site and involves meeting the precise environmental requirements of species and our ability to measure critical environmental factors such as resource availability in soils.

Restoration is intellectually challenging and requires a clear understanding of not only the nature of the ecosystem itself but also the nature of the damage inflicted and how it may be ameliorated. Bradshaw (1996) suggested that repair/restoration has two stages:

(1)   To discover and understand what is wrong
(2)   To correct the damage in an appropriate way

Sustainability is an increasingly important issue for continued economic development. During the 1970s, it was realized that there are limits to economic growth but it was not until 1987 that the so-called Brundtland Report WCED (1987) put sustainability firmly on the political agenda of the world. The idea was accepted as a political goal at the conferences of Rio de Janeiro (1992) and its successor in Johannesburg in 2002. The key to sustainability is the notion that the world should remain a suitable place to live for future generations. This is not restricted to environmental conditions alone but also has social and economic components. Jepma and Munasinghe (1998) showed the relations between the three pillars of sustainability in a triangle diagram (Figure 2.6a and b).

The National Academy of Sciences (1974) defined restoration as the replication of site conditions prior to disturbance. The term reclamation refers to rendering a site habitable to indigenous organisms, whereas rehabilitation implies that disturbed land will be returned to a form and productivity in conformity with a land-use plan. This may include creating a stable ecological state that does not contribute substantially to further environmental deterioration, which is aesthetically consistent with its surroundings.

Effective ecological restoration must involve all major levels of ecological organization from component species to entire systems. Land management practices involving re-vegetation and reclamation can be used as tools to reinitiate and accelerate succession, to enhance other natural recovery processes. This approach is viewed as 'active' restoration operating to reverse the effects of disturbance. The converse, or 'passive' restoration, is viewed as the implementation of latent management policies designed to protect impacted sites from further degradation and may involve site protection or withdrawal from use until mitigation occurs by natural succession processes start. This approach should be adopted to address both economic and ecological reasons to provide a guarantee that nature can assume a natural course following severe disturbance.

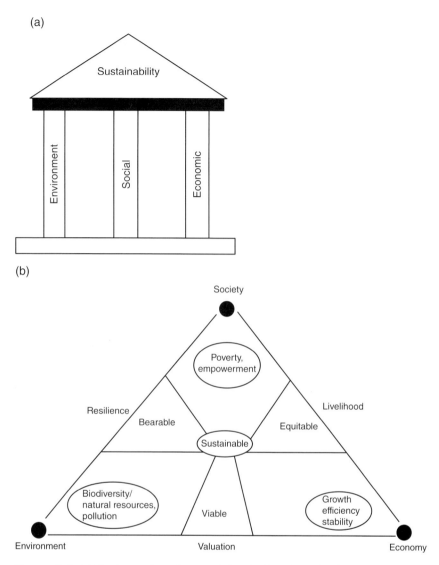

Figure 2.6 (a) The three pillars of sustainability. (b) The relationship between social, economical and environmental sustainability (Jepma and Munasinghe, 1998).

2.10 Societal ecology

Ecological restoration is necessary because the relationship between human society and natural systems is not as mutualistic as it should be. Although natural systems constitute the biological life support system of the earth, societal practices fail to realize that the elimination of species and habitats is not sustainable. It is therefore necessary to achieve

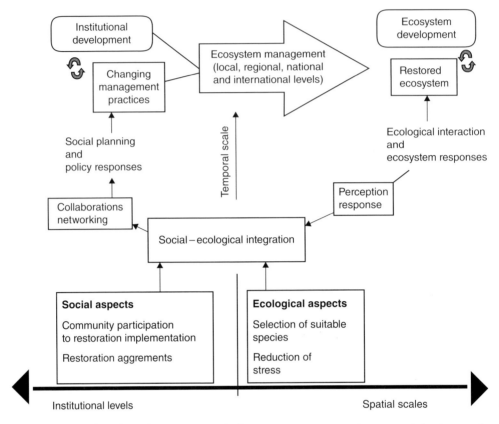

Figure 2.7 Social–ecological interactions and influences on ecosystem and institutional development for ecological management and restoration (Brunckhorst, 2010).

equilibrium between the rate of damage caused by anthropogenic pressures and the rate of restoration of ecosystems.

Societal or public ecology is a recently emerged idea to consider the increasing disparities between social, political and environmental concerns. This area of ecology links sustainability problems, community dynamics and social issues with adaptive management. Societal involvement may involve volunteer work or a multinational effort for scientific and sustainable restoration. Societal ecology, therefore, involves ecological theory and application of the scientific method along with the local communities, as they can provide several non-expert informations crucial to success. According to Brunckhorst (2010), the sustenance of ecological systems needs integration of interdependent socioecological systems across regional landscapes. Community influences to ecological restoration are reflected in new policies, resource management and scientific activities affecting restoration (Figure 2.7).

Societal ecology also deals with the formulations of how society can, and does, act on the findings of restoration ecology research programmes.

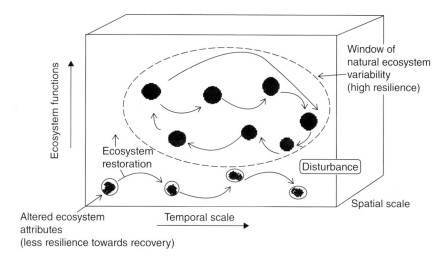

Figure 2.8 Variation in ecosystem processes over time and space (Walker and Boyer, 1993).

In the European context, societal approach of ecological restoration advocates for working beyond simple agri-environment schemes and combining them with large-scale habitat restoration. This addresses several problems simultaneously, including sea-level rise, water catchment protection, flood defence and biodiversity issues (Falk et al., 2006).

Ecological systems are highly dynamic entities as is shown in Figure 2.8 with all natural attributes, including levels of ecosystem processes (dark grey spheres), varying temporally and spatially within the window of natural variability (dashed oval line). After disturbance, the ecosystem attributes move outside the window of natural variability (mottled black spheres) and restoration is needed. However, even after 'restoration', it is unlikely for the system to attain exactly its pre-disturbance condition (Walker and Boyer, 1993).

Restoration requires multiple efforts, because multiple perturbations have pushed ecosystems beyond their ability to recover spontaneously. Full restoration means that the ecosystem is once again resilient, that is, it has a capacity to recover from stress (SERI, 2002; Walker et al., 2002). Yet, it is rarely possible to achieve the self-sustaining state because degraded ecosystems typically lack natural levels of environmental variability (Baron et al., 2002; Pedroli et al., 2002) ensuring their resilience is not recoverable (Suding et al., 2004).

Ecological restoration provides an exciting opportunity to formulate large-scale experiments and test basic ecological theory that builds the science of restoration ecology. The relationship between ecological theory, restoration ecology and ecological restoration can be viewed in a hierarchical fashion (Figure 2.9). While t`here is a very large body of ecological theory (A), only some of it can be directly applied to restoration

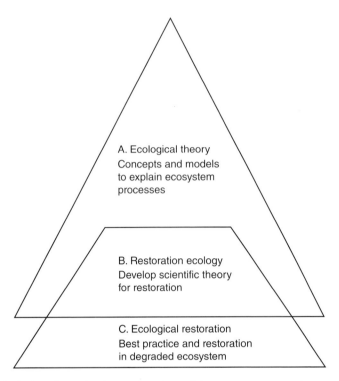

Figure 2.9 The relationship between ecological theory, restoration ecology and ecological restoration (Palmer and Bernhardt, 2006).

ecology at the present time (B). There is, thus, a demand to extend and develop theory, and the benefits of doing so extend in both directions. Ecological science benefits from the linkage, as do restoration ecology and ecological restoration. There is also a large part of ecological restoration that will never be guided by restoration ecology (C); instead, contextual constraints and societal objectives, such as co-opting natural resources or modifying ecological systems for human use, will determine restoration objectives and potential much of time.

3 Regulation of reclamation

Mining provides the building blocks for human development underpinning the need of metal and mineral products. However, the future of the mining industry is dependent on the legacy it leaves. If mines are abandoned without appropriately addressing the ecological and environmental issues, long-term detriment to the environment may occur.

The mining industry needs to fully embrace the concept of 'completion of mining' as a defined end point rather than just mine 'closure' after decommissioning has been carried out. Thus, the two important terms: 'mine closure' and 'mine completion' are used to describe the successful completion of decommissioning.

Mine closure is a process that refers to the period of time when the operational stage of a mine is ending or has ended and the final decommissioning and mine rehabilitation has been undertaken. Mine completion is the end point of mine closure, where the objective is to prevent or minimize adverse long-term detrimental environmental, physical, social and economic impacts, and the creation of a stable landform suitable for use. Poorly closed and derelict (abandoned) mines provide a legacy issue for governments and communities; and leave the mining industry with a tarnished reputation.

Mining activity often leads to negative environmental impacts, such as land degradation (opencast mining), land subsidence (underground mining), deforestation, pollution of the atmosphere and surface and subsurface waters, and the appropriateness of disposal of solid wastes, etc. Opencast mining in areas with forest cover involves deforestation and this has adverse environmental effects. The prevention and mitigation of these effects in the longer-term via re-vegetation and the repair of affected areas form an integral part of mine development.

Reclamation of Mine-Impacted Land for Ecosystem Recovery, First Edition. Nimisha Tripathi, Raj Shekhar Singh and Colin D. Hills.
© 2016 John Wiley & Sons, Ltd. Published 2016 by John Wiley & Sons, Ltd.

Salient features regarding mine closure and rehabilitation are:

- Concurrent rehabilitation of mined-out areas with ongoing extraction
- Rehabilitation of the mined-out areas in a phased manner
- Appropriate (safe) disposal of by-products and waste; and
- Implementation of an agreed environmental management plan (EMP) and a time-led reclamation program for minimising environmental harm and full-restoration of the mined-out areas.

Several countries have their mining laws and regulations coordinated with environmental standards for air, water, noise and dust emissions. However, the main objective of mining policies should be aimed at recognising that it is possible to anticipate the legacy of a mining at its conception. This includes: planning closure, the identification of economic impact and the determination of final land-use objectives. This process should involve consultation with the local community, to ensure progressive rehabilitation and the needs of the community that is affected by mining and how these will be reconciled following closure.

Very few countries have legislation addressing the management of the reclamation process, although some require a financial provision (a fund) or a reclamation 'bond' (Intaraprvich and Clark, 1994). A reclamation bond is used to assure that mining companies reclaim the damaged land and manage water resources specific to mining, whereas the 'fund' approach can be used to finance the environmental management reclamation of the mine site.

3.2   Mining laws and policies in India

Land degradation is the most serious impact from surface mining. Apart from polluting air and water, opencast or surface mining can leave land unsuitable for other uses. Some environmental problems include severe physical degradation of the surrounding landscape and despoliation of water resources, which impacts on agricultural land and low-lying habitats. Therefore, to safeguard existing habitats and to minimize environmental impact the government of India is responsible for the enforcement of regulations related to mining, and land reclamation and rehabilitation.

3.2.1   Status of legislation for land reclamation

No legislation has been enacted in India to date to enforce land reclamation after completion of mining operations. Although Section 27(2) of Mineral Concession Rules (MCR) (1960), bestowed to the State Government, lays down conditions for working of a mineral deposit, no State Government has incorporated the conditions to make land reclamation necessary after completion of the mining operations.

The key features of various Mines and Minerals Policies in India have been summarized in Table 3.1.

3.2.2 National Mineral Policy: environmental protection

The extraction and development of minerals are interlinked with the management of natural resources including: land, water, air and forests (Singh, et al. 1994). The areas in which minerals occur often have other resources-utilization issues. Some areas are ecologically fragile and others are biologically rich and so it is necessary to have a comprehensive approach towards development and the protection of the environment. Thus, through a co-ordinated approach the sustainable development of mineral resources can take place with the minimum of impact on the environment (Gonzalez-Perez and Leonard, 2013).

3.2.3 The Mines and Minerals (Regulation and Development) – MMRD Act, 1957

Under the Mines and Minerals (Regulation & Development) – MMRD Act, 1957 (Amended in 1986 and 1994), the Central Government of India was not empowered to prevent or control of environmental degradation resulting from mining. This act addressed this by ensuring the environmental management of the mining areas, with certain amendments proposed to the MMRD Act, 1957.

Under MM (R&D) Amendment Act, 1999, the Act was renamed as Mines and Minerals (Development and Regulation) Act 1957 (Draft IBM Manual on Appraisal of Mining Plan, 2013).

3.2.4 The Mineral Concession Rules, 1960

The MCRs were constituted under MMRD Act, 1957. These direct all mining plans to include mitigation measures and their implementation.

3.2.5 Mineral Conservation and Development Rule, 1988

The MCDR was reorganized in 1988 to incorporate provisions on submission of mining plans, reclamation and rehabilitation of land, restoration of flora, and environmental protection. As as result all metalliferous mines are required to submit an environmental management plan to the Indian Bureau of Mines (IBM), as an integral part of the mining plan. There is no equivalent/comparable Act/Rule for coal mines at present. The IBM circular emphasizes the reclamation planning for the prospective mine land over the next 5 years at the time of excavation planning.

Table 3.1 Mining and minerals policies in India.

Sl	Policy/act/ rule	Purpose	Key aspects	Responsibilities
1	National Mineral Policy, 1993	Mineral Development and Protection of Environment	Mineral(s) processing and repairing affected forest area shall form integral part of mine development strategy in accordance with the prescribed norms	Avoidance of mining operations in the forest area
2	The Mines and Minerals (Regulation and Development) – MMRD Act, 1957	Systematic development and conservation of mineral resources	• Section 18 – Empowers Central Government to frame rules for mineral conservation and development • Section 4A – Premature termination of the mining lease in the interest of regulation of mines and mineral development – No premature termination possible to check damage to ecology and environment • Amended Section 4A – Confers the Central Government power to prematurely terminate the mining lease to check damage to ecology and environment • Amended Section 5(2) – No mining lease shall be granted by State Government unless it ensures the – Evidence of leased area being mined earlier with established mineral(s) existence – Existence of mining plan for development of mineral deposits, approved by Central/State Government	Mineral conservation and development
3	Mines and Minerals Development and Regulation Amendment Bill, 2015	Rehabilitation and welfare of people affected due to mining activity	• Aid in streamlining the process of grant and renewal of mining leases of commodities like iron ore, manganese ore, bauxite, etc. • Contribution to District Mineral Fund (DMF)	Rehabilitation
4	The Mineral Concession Rules, 1960	Environmental protection	• Mining plans to incorporate natural water course, forest areas, density of trees and other important natural resources • Impact of mining on forest, land surface, air and water • Rule 22(3) (e) (iv) – The mining plan should consider land reclamation and pollution control • Amended Rule 58 – Empowers Central Government to reserve vacant area for purely conservation purpose for an indefinite period	Land reclamation, ecological/resource conservation

5	Mineral Conservation and Development Rule (MCDR), 1988	Conservation and development of the minerals	• Conferred by Section 18 of the MM (D&R) Act, 1957
			• Cancelled the MCDR – 1958
			• Applies only to metalliferous mines (excluding atomic and minor minerals)
			• Rule 23 – Deals with abandonment of mines
			  – Submission of 90 days prior abandonment notice by mines to IBM
			  – Protection of abandoned mine workings and the environment
			• Rule 34 –The provision for a year-wise afforestation programme for the first 5 years and a conceptual plan for the remaining part of the life of the mine
			• Rule 31 (Chapter V) – Environmental protection and pollution control
			• Rule 32 (Chapter V) – Removal and utilization of topsoil for restoration/rehabilitation or storage for future use
			• Rule 33 (Chapter V) – Related to storage of overburden and waste rock
			  – Storage of overburden/waste rock/rejects/fines over separate dumps
			  – Spoil dump management to check material run-off
			  – Impervious ground for dumps/tailings to prevent leaching due to precipitation
			  – Back filling with waste rocks and overburden, wherever possible
			  – Terracing of waste dumps and stabilization with vegetation, where backfilling is not possible
			  – Safe deposition/disposal of fines/rejects/tailings to check flow in agriculture field and water bodies
			• Rule 34 (Chapter V) – Deals with reclamation and rehabilitation of affected lands
			  – Phased restoration/reclamation/rehabilitation before the mine completion and abandonment
			• Rule 40 (Chapter V) – Deals with flora restoration
			  – Least damage to flora
			  – Plantation not less than double of trees destroyed by mining
			  – After care of trees during the lease; thereafter handed over to the forest department
			  – Relevant statutes for permissible limits of pollutants, toxins and noise
				• Submission of environmental management plan by metalliferous mines to Indian Bureau of Mines (IBM)
				• Precautions for environmental protection and pollution control while conducting prospecting, mining, beneficiation/ metallurgical operations
				• Least damage to flora and restoration/ reclamation/rehabilitation of affected lands

6	Forest Conservation Rule, 1980	To check deforestation	• No use of forest land for any non-forest purpose (including mining) will be permitted without the prior approval of the Central Government	Compulsory reforestation

(continued)

Table 3.1 (Continued)

7	National Forest Policy, 1988	Environmental stability, ecological balance and atmospheric equilibrium for the sustenance of all life forms	• Repairing and revegetation of forest land used for mining and quarrying • No provision of lease grants for mining without a proper mine management plan considering environmental perspectives and enforced by adequate machinery • A detailed management plan ensuring repair of damaged ecosystems
8	Policy Statement on Abatement of Pollution, Environment and Development, 1992	Evolve mechanism to reduce local concentrations of pollutants at complex industrial sites	• Identification of 'critically polluted' areas • Regional Environment Management Plan for these areas • Formulation of source-related Mass-Based Standards • Standards would not merely be a regulatory tool but would also be a mechanism to promote technological upgrading to prevent pollution • Evolution of codes of practice and guidelines for specific purposes
9	National Mineral Policy, 1993	Minimize adverse effects of mineral development on forest ecology and environment through appropriate protective measures	• Para 7.13.1 – Some areas with minerals are ecologically fragile and biologically rich. The policy, to some extent, recognized the environmental concerns and stated that the mining operations shall not be taken up in identified ecologically fragile and biologically rich area • Para 7.15 – Deals especially with the mining activities spread over a few decades for the orderly and systematic mine closures and rehabilitation. However, it does not deal with the environmental aspects of the closure of mine • A comprehensive survey to facilitate the choice of land use, with consideration of development needs and protection of forest, the environment and ecology • Sustainable development of the mineral industry in harmony with environment • Reclamation and afforestation of appropriate land concurrently with mining/mineral extraction • Conversion of old abandoned mine sites into forests or other appropriate land uses
10	National Conservation Strategy and Policy Statement on Environment and Development	Enactment of laws for appropriate land uses	• To protect the soil from erosion, pollution and degradation • Restoration and reclamation of mined out areas • Implementation of the environmental management plans during mining to ensure adequate ecological restoration of the affected areas • Rehabilitation of abandoned, mined out areas in a phased manner so that the land resource can be brought back under productive use • Provision of mining leases stipulating tenure, size, shape and disposition with reference to geological boundaries and other mining conditions to ensure systematic extraction of minerals along with environmental conditions

Although there is provision for abandonment of mines, no precise definition is given, and as such confusion arises between the closure and an abandonment of a mine.

3.2.6 Forest Conservation Rule, 1980

Afforestation is one of the important approvals condition as proposals for the diversion of forested land for non-forest use are required. Compensatory afforestation must be carried out over equivalent area (of that removed).

3.2.7 National Forest Policy, 1988

There is a need to strike a balance between two activities of national importance, i.e. mining and forest conservation and this should be reflected in policy. According to the available statistics, the land area covered by mineral reserves is only about 1% of the total of the country (Ministry of Environment and Forests, 2008). Since the forested government land is released to mining project on a lease-basis, it is reverted back to the Forest Department after mining has ceased.

3.2.8 Policy Statement on Abatement of Pollution, Environment and Development, 1992

As specified in the policy, this mechanism would reduce local concentrations of pollutants in complex industrial sites. Only fragmented work has been carried out without follow up action in the critically polluted areas.

The specifications of source-related Mass Based Standards require the most polluting industrial processes and application of best available demonstrable technological (BDAT) solutions. However, for mining, BDAT remains poorly developed.

The policy is aimed at the cost of using environmental and natural resources, and thus, provides fiscal incentives for the installation of pollution control systems.

Economic instruments are also promoted to internalize the cost of pollution and conserve resources, particularly water. Effluent charges are proposed based on the nature and volume of the releases to the environment to promote optimization, however, this has not yet been put into practice.

3.2.9 National Mineral Policy, 1993

The adverse effects of mineral development on forest ecology and the wider environment are addressed through both appropriate protective and 'preventive' measures during mining.

The NMP contains guidelines to ensure the enactment of legislation towards the impact of pollution from mining, and placement of measures for environmental protection and ecological restoration. However, not all these instruments contained have been legislated so far.

The policy recognizes environmental impacts from mining and restricts mining in both ecologically fragile and biologically rich area. However, the mining of scarce minerals has been exempted.

3.2.10 National Conservation Strategy and Policy Statement on Environment and Development

In India, out of the total of 329 Mha, 175 Mha of land requires special treatment for restoration. Degradation is caused by water and wind erosion (150 Mha), salinity and alkalinity (8 Mha) and river action and other factors (7 Mha). The policy stresses the appropriate use of land to protect the soil from erosion, pollution and degradation, and for the restoration and reclamation of mined out areas.

3.2.11 Environmental auditing and accounting of geographical resources excavated from mines

The exploitation of geological and biological resources involves environmental auditing and accounting (Tietenberg, 1998). The environmental impact of extraction and processing of minerals tends to be much greater when a forest cover is removed, as the permanent loss of biological diversity and the extinction of elusive plant and animal species may result. Other important factors associated with destruction of trees include the loss of carbon dioxide absorbance (by trees) and the production of oxygen (via photosynthesis). In addition, the germ plasm of disease resistant genes found in the wild relatives of valued crop plants, is impacted, as the destruction of forest and the vegetative cover takes place.

The level of ecological destruction due to mining is presented in Table 3.2.

3.2.12 EMP for sustainable mining activities

Economic growth and development depend not only on resource management, but also on sustainable resource utilization. This aspect of industrialization has not been addressed by environmental planning in the past. As the past 20 years have seen significant population

Table 3.2 Ecological destruction caused by extraction of mineral ores from mines.

	Activity	Potential impact
1	Excavation and ore removal	i. Destruction of forest; loss of habitats and biodiversity
		ii. Soil erosion and overburden
		iii. Land subsidence
		iv. Loss of topsoil and microbes
		v. Enhanced carbon foot print
		vi. Crippling effects on workers
		vii. Occupational disease, including silicosis and tuberculosis.
2	Ore concentration	i. Waste generations (tailings)
		ii. Organic chemical contamination
		iii. Acid drainage
		iv. Water pollution
3	Smelting	Air pollution, from e.g. emission of arsenic, lead, cadmium and other priority contaminants including fluoride, hexavalent chromium, iron, copper, etc.

growth and an increased demand on resources, the uncontrolled exploitation of minerals has required new management tools. Until recently the impact of mining-related pollution in India has been underrated, but the introduction of environmental management plans (EMP) for sustainable mining practices ensure future harm is minimized. An EMP is the first step, in quantifying the extent of reserves of the deposit, and its coordinated extraction and processing in a systematic and sustainable way.

3.2.13   Displacement in the mining sector

Mining is a site-specific operation, where the mineral of interest can occur in horizontal layers that follow the land surface. However, as the structural geology of deposits can be complex, mining companies can face high levels of uncertainty around future their land requirements (Owen and Kemp, 2015), and this leads to enhanced human and habitat displacement. In India, mining displaced 2.55 million people between 1950 and 1990 (Downing, 2002).

The problem of mining-induced human displacement poses major risks to societal sustainability and wellbeing. Displacement involves the physical eviction from a dwelling and also the seizing of productive land (Cernea, 2000).

However, during mining planning, there is clear objective to minimize displacement, or to resettle or rehabilitate if displacement proves inevitable. In underground mining, displacement can be minimized by adopting the process of stowing underground voids. However, during opencast mining the magnitude of displacement partially depends on geological factors. Resettlement and Rehabilitation (R&R) Policies were first directed in the 1990's at coal mining, being published in 1994, and updated in 2012. Specifically, India's coal-rich states introduced their own policies (e.g., Orissa in 2006 and Jharkhand in 2008) (Gopalakrishnan, 2006).

The primary requirement of R&R Policy is to minimize land used, by ensuring planning of a mine and ancillary units are minimized and time limited. This warrants analysis of the alternative options for land acquisition during the planning process, including utilizing wasteland/poor quality land rather than cultivated or agricultural land.

To repair the large-scale displacement that has occurred in the mining sector, resettlement and rehabilitation have become central issues of the mine developmental process. The R&R Policy ensures that affected people have improved well-being (especially those who are below the poverty line), or as a minimum regain their former standard of living, and this may include provision of housing and community infrastructure within a specific time-frame.

The important issues to be addressed and ensured via the R&R are:

■ Compensation benefits should not be limited to the land/immovable property holders but also to the weaker section of the society. Special attempts are required to ensure that women are given adequate access to income-generating opportunities offered under the policy.
■ Full responsibility of all costs and actions necessary to successfully resettle and rehabilitate the directly affected people in terms of livelihood, housing, infrastructure and access to services.
■ The compensation policy should aim that the affected people achieve at least a reasonable standard of living within a specified period of time. The compensation should be assessed based on official value of land and wherever the official figures are not available, the estimation should be based on the current official guidelines.
■ Assistance via imparting vocational training to people to settle and develop in different trades.
■ If opportunity arises with the mining project, the affected people should be given priority to get benefits.
■ Provision of incentive for reasonable period, for timely evacuation of proposed mine areas.
■ Minimal impact on the existing social fabric of the community, such that community facilities including places for worship, playgrounds, etc. are not overlooked during resettlement.

However, displacement by mining generates false claims for the compensation, and effective mechanism to protect a mining project from fraud/unjustified claims must be implemented.

3.2.13.1   *R&R policies*

Mining is a site-specific operation and minerals are extracted where they are found. Thus, there is very limited flexibility in the location of mining projects. However, the associated infrastructure, including materials handling, beneficiation and other support-services may be located to ensure avoidance of sensitive areas.

To ensure the rehabilitation and resettlement in mining areas, the following legislative Acts are in operation in India (Table 3.3).

The compensation package under the new Act (2013) ensures that claimants will be entitled to compensation, which is determined on the basis of the market value of land, determined as on the date of preliminary notification. According to Section 34, if there is delay in payment of compensation beyond 1 year from the date on which possession is taken, interest at the rate of 15% pa shall be payable from the date of expiry of the said period of 1 year on the outstanding amount of compensation until the date of payment.

The Act also stipulates mandatory consent of at least 70% of affected people for acquiring land for Public Private Partnership (PPP) projects and 80% for acquiring land for private companies (http://timesofindia.indiatimes.com/india/New-land-acqui sition-law-comes-into-force/arti cleshow/28204302.cms).

Table 3.3 Legislation for rehabilitation and resettlement.

	Act	Purpose	Key aspects
1	Land Acquisition Act (1894)	Address the effect of mining in respect to • Displacement of population • Loss of agricultural land • Loss of livelihood and household properties	• Rehabilitation of affected people to minimize R&R conflict • Unless R&R is addressed, the project implementation may be delayed
2	The Right of Fair Compensation and Transparency in Land Acquisition, Rehabilitation and Resettlement Act (2013)	Relating to land acquisition, compensation, rehabilitation and resettlement	• Replacement of Land Acquisition Act, 1894 • Fair compensation to those whose land is taken away • Transparency to the process of acquisition of land to set up factories or buildings, infrastructural projects • Regulations for land acquisition as a part of India's massive industrialization drive driven by public–private partnership

3.2.14 Gaps in mining and environmental legislation and recommendations

The Constitution of India enunciates Directive Principles of State Policy which is stated as: 'The State shall strive to promote the welfare of the people by securing and protecting as effectively as it may a social order in which justice, social, economic and political, shall inform all the institutions of the national life'. The Directive Principle has been subsequently enlarged vide Article 48 A (inserted in 1976) and Article 49. Additionally, to deal with the environmental issues, Article 51(A) has also been incorporated in the India Constitution. These national sectorial policies are governed by the principles and directions given in the Constitution of India.

Considering the changing global economic scenario and hence the emerging aspirations and needs of the society, the policy of the government needs to be dynamic and also assuring the strategic interests of the country. Acts and Rules are formulated and amended from time to time to implement the stated policies.

In India, most mining environment-related policy/policy statements were formulated between 1980 and 1994. To meet the requirement laid down, amendments are made progressively. Despite the fact that there are environment-related policies/policy statements supported by appropriate instruments, the country has not achieved the desired level of environmental protection in the mining sector.

Many coal-mining areas in India (e.g., Jharia, Raniganj and Bijolia) have still been not incorporated reclamation in their environmental management plans, and this is especially important where the management of top soil and spoil involve slope stability, runoff and long-term impact-related issues.

3.3 International policies and legislations

There are many differences in way the national and regional governments of Australia, Canada, UK, Europe, Japan and the United States, deal with the specific aspects of mine closure and its management thereafter. The greatest differences are in the level of specific policy and legislation that is in place for the abandonment and post-closure responsibilities of existing and planned mines (Table 3.4). For these, the subordinate levels of government in Australia and Europe have each evolved fairly consistent and comprehensive policies and legislation, whereas, in Canada and the United States, there is considerable variation among and between the subordinate levels of government (www.elaw.org/system/files/11198931391clark_jcclark.pdf).

Table 3.4 Legislative provisions for mine closure in the provinces/territories of Australia and Canada, Europe and individual States of the United States.

Country, state or province	Specific provisions for reclamation	EIA required before lease	Bonding procedure	Provisions for Abandonment	Provision for Noncompliance
Asia					
Japan	×	×	×	×	×
Australia					
New South Wales	×	×	×	—	×
Northern Territory	×	×	×	—	×
Queensland	×	×	×	×	×
South Australia	×	×	×	×	×
Victoria	×	×	×	×	×
Western Australia	×	×	×	×	×
Canada					
British Columbia	×	×	×	×	×
Manitoba	×	—	×	—	×
New Brunswick	×	×	×	—	—
Northwest Territories	×	×	×	—	×
Nova Scotia	×	×	×	×	×
Ontario	×	×	×	×	×
Quebec	×	×	—	—	—
Saskatchewan	×	×	—	—	—
Yukon Territory	×	×	×	—	×
Europe					
Germany	×	×	×	—	×
Ireland	×	×	×	×	×
United Kingdom	×	×	×	×	×
Wales	×	×	×	×	×
United States					
Alaska	×	×	×	—	×
Arizona	×	×	—	—	×
California	×	×	—	×	×
Montana	×	×	×	—	×
Nevada	×	×	×	—	—
New Mexico	×	×	—	—	×
Utah	×	×	—	×	×
Washington	×	×	—	—	×
Wyoming	×	×	—	×	×

Sources: Fortin (1992), Gallaher and Lynn (1989) and Intarapravich and Clark (1994).

3.3.1 Mine reclamation laws and policies in the United States

The mining of minerals, including gold, uranium and other metals in hard rock on public land, is governed by the General Mining Law of 1872. The General Mining Act of 1872 authorizes the prospecting and mining of economic minerals on federal public lands. This law, approved on 10 May 1872, was formulated to protect mining on public land in California and Nevada from the late 1840s through the 1860s, including during the California Gold Rush. All citizens of the United States 18 years or older have the right under this law to locate a lode (hard rock) or placer (gravel) mining claim on federal lands open to mineral entry.

In the United States, the Surface Mining Control and Reclamation Act (SMCRA) was enacted in 1977, after the US Congress recognized the need to regulate mining activity, rehabilitate abandoned mines and protect society and the environment from the adverse effects of mining operations. The essential components of the SMCRA are described in Table 3.5.

The SMCRA did not prohibit mountain top coal mining, an activity that steadily increased after 1977 (Kenney, 2007). The general provisions of SMCR Act are as follows:

(a) The extraction of minerals is essential to the economic well-being of the state and to the needs of the society.

Table 3.5 The essential components of Surface Mining Control and Reclamation Act, 1977.

Sl	Act	Purpose	Key aspects	Responsibilities
1	SMCR Act (1977)	• To permit guidelines for existing and future mines	• A comprehensive surface mining and reclamation policy	States regulate the environmental effects of coal mining
	SMCRA, Public Resources Code, Sections 2710–2796	• Establish a trust fund to finance the reclamation of abandoned mines	• Regulation of surface mining operations to assure the minimized adverse environmental impacts and reclamation of mined lands to a usable condition • Encourages the production, conservation and protection of the state's mineral resources	The Federal Government retains oversight responsibility carried out by OSM
a	Public Resources Code Section 2207		• Annual reporting for all mines in the state • Chapter 9, Division 2 – State Mining and Geology Board is required to adopt State policy for the reclamation of mined lands and the conservation of mineral resources	State Mining and Geology Board is also granted authority and obligations

(b) The reclamation of mined lands is necessary to prevent or mini-mize adverse environmental effects and to protect the public health and safety.

(c) The reclamation of mined lands will permit the continued mining while ensuring the protection and subsequent beneficial use of the land.

(d) The production and development of mineral resources contribut-ing to the state's strong economy are vital to reducing transporta-tion emissions.

(e) The need to provide: local government, metropolitan planning organizations and other relevant planning agencies with the infor-mation necessary to identify and protect mineral resources within general plans.

(f) The state's mineral resources are vital, finite and important natural resources, and the responsible protection and development of these mineral resources are vital to a sustainable state.

The SMCRA is widely seen as a highly effective piece of legislation; it is unlikely that it will be used as a model for future regulations govern-ing hard rock mine reclamation (CAMMA, 2001). This is because the SMCRA is expensive to administer, and unlike the US coal industry, these costs cannot easily be passed on to the consumer. In this way, the US mining regulations are based on a patchwork of laws of varying effec-tiveness and severity governing how hard rock mines will be reclaimed. In the absence of any standardized laws that are specifically written to address reclamation, the emphasis will remain with the mining industry to develop and follow best management practices if the existing situation is to substantially improve (Campusano and Patricio, 2001).

In the State of California, there are a number of different agencies administering their own laws and regulations that apply to mine clo-sure. The two most important are the local and the state agencies (Table 3.6).

Table 3.6 Key agencies and their roles in the State of California.

Sl	State agencies	Purpose	Key aspects
1	Local agencies	Control land-use law	• Regulate post mining land use • Assure that the final topography, drainage, re-soiling, re-vegetation, etc. are appropriate to support the post mining land use
2	State agency	Responsible for water quality protection	• The Water Quality Control Board has jurisdiction over the 'mine waste units' • Important consideration are waste rock dumps, tailings ponds, spent heaps, etc.

The land-use plans in the United States are prepared for all public lands. Uses of these lands by facilities such as mining operations must conform to the land-use plan and its stipulations on particular uses of the land (Bureau of Land Management, 2001).

3.3.1.1   Intent of the legislature

The intent of the legislature includes the creation and maintenance of an effective and comprehensive surface mining and reclamation policy with regulation of surface mining operations so as to assure that:

(a) Adverse environmental effects are prevented or minimized and that mined lands are reclaimed to a usable condition, which is readily adaptable for alternative land uses.
(b) The production and conservation of minerals are encouraged, with consideration to values relating to recreation, watershed, wildlife, range and forage and aesthetic enjoyment.
(c) Residual hazards to the public health and safety are eliminated.

3.3.1.2   State policy for the reclamation of mined lands

State policy shall apply to the conduct of surface mining operations and shall include, but shall not be limited to, measures to be employed by lead agencies in specifying grading, backfilling, re-soiling, re-vegetation, soil compaction and other reclamation requirements, and for soil erosion control, water quality and watershed control, waste disposal and flood control.

The American states which mine both coal and non-coal resources have two separate sets of requirements. The state's reclamation programme for non-coal resources is not subject to any national oversight, and states are free to construct their own requirements. Thus, in every state, there are at least two sets of regulations. It is not uncommon for states to divide regulatory responsibility further and have more than two separate regulatory schemes. For example:

Colorado:
- Colorado Land Reclamation Act (in addition to a Coal Mine Reclamation Act), for the extraction of construction materials (such as sand, gravel, stone, borrow material)
- Colorado Mined Land Reclamation Act, which applies to all other forms of mining, including metal, or 'hard rock' mining

Montana:
- Opencut (opencast) Mining Act
- Specific permits for cyanide use

Idaho:

- Ordinary permit system
- Special permits for placer/dredge operations and for cyanide operations

Arizona: Still has a unitary system for implementation of its law, which is still in development. The law applies to surface disturbances of five acres (about two hectares) or greater and only to metal mining. Under the state's new Mined Land Reclamation Act, all new metal mining operations, whether in the exploration phase or the production phase and which meet the size requirement, are required to obtain approval of reclamation plans and financial assurances (Danielson and Nixon, 2000).

3.3.1.3 *National Mining and Minerals Policy*

The National Mining and Minerals Policy is a continuing policy of the Federal Government in the national interest to foster and encourage private enterprise in the:

- Development of economically sound and stable domestic mining, minerals, metal and mineral reclamation industries
- Orderly and economic development of domestic mineral resources, reserves and reclamation of metals/minerals to ensure satisfaction of industrial, security and environmental needs
- Mining, mineral and metallurgical research (including the use and recycling of scrap) to promote the efficient use of natural and reclaimable mineral resources
- Study and development of methods for the disposal, control and reclamation of mineral waste products and the reclamation of mined land, so as to minimize any adverse impact of mineral extraction and processing upon the physical environment

The 'minerals' mentioned in the policy shall include all minerals and mineral fuels including oil, gas, coal, oil shale and uranium.

3.3.2 Mining laws and policies in the United Kingdom

In Britain, mine legislation is enacted by Acts of Parliament, which are further refined by Regulations. These Regulations explain, extend or amend the parameters applicable to the relevant Act.

In Britain, the major mines disaster led to the formulation of specific legislation to control or eliminate certain operation within the mining industry. The early legislation in the nineteenth century provided for a safer working environment with controlled ventilation. With the growth of mining, additional legislation was introduced to ensure the provision of working plans, mine abandonment plans and the creation of a Mines Inspectorate, to enforce the regulations. However, until recently, mine legislation was mainly concerned with the working environment of quarry without considering the environmental impact of mining.

Coal mining in Britain has been associated with subsidence, waste tip instability and pollution. Considering these problems, especially, since the Aberfan disaster (1966), the environment has been added to the legislation via the Mines and Quarries (Tips) Act 1969, governing mines and quarries and their attendant waste tips. This Act states that it is 'An Act to make further provision in relation to tips associated with mines and quarries; to prevent disused tips constituting a danger to members of the public; and for purposes connected with those matters'. This Act is an extension of The Mines and Quarries Act 1954 which did not mention tips specifically in its provisions.

The implementation and compliance requirements of the Mines and Quarries (Tips) Act 1969 are laid out in The Mines and Quarries (Tips) Regulations, 1971. Subsequently, the Quarries Regulations 1999 state that tips must be designed, constructed, operated and maintained so that instability or movement likely to cause risk to the health and safety of any person is avoided. They also specify the geotechnical and other measures to be taken to ensure this. Other legislation may have some bearing on the construction, operation and disposal of mineral waste tips, including the Rivers (Prevention of Pollution) Acts and the Clean Air Acts.

Gold and silver mines in Britain are known as 'Mines Royal', and the Crown holds the rights to these metals. This is the case across the whole of the United Kingdom; except in Scotland, where these rights have been transferred by charter. The exploration and development of gold and silver require the licence, known as the Royal Mines Licence, obtained from the Crown Estate Mineral Agent, Wardell Armstrong.

Other metallic and industrial minerals in Britain are in private ownership, and although there is no national licensing system for exploration and extraction, planning permission must be obtained from a mineral planning authority for their extraction.

3.3.3   Mining laws and policies in European Union

Major challenges for the mining industry include competition for land use, a heavy licensing process and compliance with stringent environmental laws. The European Union (EU) legislation is vast because of applicable regulatory safety and environmental demands on mining activities in EU member states. In addition to general industrial regulation, there is specific mining legislation in the EU, which considers the environmental impact of mining (especially waste and groundwater), as well as occupational health and safety. However, it neither contains a general regulation of mining as such nor mining safety, and therefore, each member state's specific legislation governs mining safety standards in the EU (BRGM, 2001). Nevertheless, Sweden and Finland, different from the rest of the EU, have an active mining industry, and new mining safety legislation has been enacted in these member countries. In new Finnish Mining Act bill, mine safety is one important key area.

In EU member states, the mining technologies especially focus on finding further resource deposits in established mines, economically and environmentally viable exploration, processing of ores with low levels of mineral or metal concentration and reducing mining's surface footprint.

However, the mining activities in EU member countries result into one of the largest waste streams. Mining operations raise the following environmental concerns:

■ Depletion of non-renewable resources
■ Air, soil, water and noise pollution
■ Natural habitat destruction/depletion
■ Visual impact on the landscape and effects on groundwater levels

3.3.3.1 *Legislative framework for the safe management of mining waste*

The legislative framework in the EU includes the directions, which ensure the safe management of mining waste:

■ The Mining Waste Directive: It includes obligatory permits and setting requirements for building or modifying an extractive waste facility. If potential risk to the environment or public health exists, operators need to provide a financial guarantee and draw up emergency plans, a policy for prevention of major accidents, and develop safety management systems.
■ The Best Available Techniques (BAT) document: It documents the management of waste from ore processing (tailing) and waste rock in mining and also promotes activities considered as 'good practice'.
■ An amendment of the Seveso II Directive: It covers the risks arising from storage and processing activities in mining, particularly tailing ponds and dams used in mineral processing of ores.

3.3.4 Mining laws and policies in Australia

Mining is a significant component of Australia's economy and was a key factor in assisting the country through the global financial crisis. In Australia, mining development is administered by the states and territories, but the Commonwealth (Federal) Government's Environment

Protection and Biodiversity Conservation Act 1999 (effective since July 2000) has established a nationally consistent framework for environmental assessment (EA) of new projects and variations to existing projects, based on consultative agreements between the two levels of government.

In Australia, liability for environmental protection during the operational phase of mining lies with the leaseholder or landowner in most cases. At the time of closure, the mining lease is relinquished or extinguished and the landowner resumes liability. Each state has specific rules though (Clark, 1999):

- Tasmania: The Crown is liable after the discharge of corporate liability with the discharge of lease and closeout.
- South Australia: Landowner is responsible, unless another arrangement has been agreed at closeout.
- Western Australia: Land title-holder or the Crown.
- Queensland: Liability is with the landowner, but a mining company may be held liable under the Conservation Act if subsequent problems arise.
- New South Wales: In the case of coal, the mining leaseholder remains liable. In the case of minerals, the landowner is liable.
- Victoria: Liability rests with the landowner, but the mining company can be prosecuted if problems related to mining arise and additional cleanup is required.
- Northern Territories: Once the mining company has quit the site, liability rests with the land title-holder.

3.3.5   Mining laws and policies in Canada

The Federal Government of Canada is responsible for mine closure and mine reclamation in the Yukon Territory, Nunavut and the Northwest Territories. Minerals and metals activities throughout the rest of Canada (excluding some uranium mines) are managed and regulated by the provincial governments.

In Canada, policy framework continues to evolve:

- Devolution of federal regulatory responsibilities to regions and territories
- Aboriginal control over lands and resources
- New land planning and EA regimes

In this regard, policy must be based on both scientific rigour and adaptive approaches with consideration of socio-economic and cultural elements (Clark, 1996). From a policy perspective, it is possible to reproduce the Canadian government priorities, distinguishing between existing and new mines. Key priorities regarding new mines are the following:

■ To develop reclamation and decommissioning standards those are in keeping with other standards in Canada and elsewhere.
■ The EA process must cover closure options, processing and ongoing reclamation.
■ Appropriate terms and conditions for site reclamation and decommissioning, through permits and licensing procedures.
■ To ensure that closure plans are updated and that sufficient financial security (bonds, assurances, etc.) in place prior to development.
■ To effectively conduct inspection and enforcement.

Key priorities regarding existing mines are the following:

■ When a mine operator has become insolvent or is unable to finance the costs of reclamation, responsibilities revert to the Crown.
■ If ownership has reverted to the Crown, the government must conduct an independent EA and site-decommissioning plan.
■ Risk assessment, both financial and physical, must be completed.
■ Procuring financial resources: targeting past owners through waste management regulations, or joint public/private funding programmes.

3.3.6   Mining laws and policies in South Africa

Legislation in South Africa governing mining has been in existence for many years. The mining laws periodically undergo review and amendment with the last major change in 1991, which consolidated a number of different laws dealing with precious metals, diamonds and base minerals (Mining, Minerals and Sustainable Development Report, 2002).

One important consideration for such legislation is that any decisions on the closure requirements and whether proper closure has taken place should be done cooperatively within government, for example, by an intergovernmental institution including representatives of government/state departments who have responsibility for the protection of the environment, water soil, etc. and social issues (Dirección General de Asuntos Ambientales, 1999). The goal is to ensure that there is no conflict between the different interest groups. Moreover, the government approach is a 'risk zero' strategy. The idea is that any new permit and extension of an already approved project would be required to provide funding such that, at any given time during the life of a mine, there should be sufficient money available to meet the environmental liabilities should the mine suddenly close down (Parsons and Kilani, 2000).

3.3.7   Mining laws and policies in Sweden

The Swedish Parliament has merged the Mining Act, 342 (1974) and the Act concerning Certain Mineral Deposits, 890 (1974) into the Minerals' Act, 45 (1991), which became effective from 1 July 1992. The Act is applicable to exploration and exploitation on land irrespective of the ownership.

Over the last century, Sweden has had the following laws relating to minerals:

- The Mining Regulation Act 24 (1884), which was replaced by
- The Mining Act 314 (1938), which was in turn superseded by
- The Mining Act 342 (1974)
- The Coal Deposits Act 46 (1886)
    - The Uranium Act 679 (1960), with the two above being replaced by:
- Certain Mineral Deposits Act 890 (1974)

The Minerals Act currently in force replaced both the Mining Act and the Act concerning Certain Mineral Deposits of the same year, 1974.

The Minerals Act applies to a number of different minerals, for example, lead, gold, copper, silver, uranium, oil, gaseous hydrocarbons and diamond. The major mineral substances (concession minerals) covered by the Act are:

- Antimony, arsenic, beryllium, bismuth, caesium, chromium, cobalt, copper, gold, iridium, iron occurring in the bedrock, lanthanum and lanthanide series, lead, lithium, manganese, mercury, molybdenum, nickel, niobium, osmium, palladium, platinum, rhodium, rubidium, ruthenium, scandium, silver, strontium, tantalum, thorium, tin, titanium, tungsten, uranium, vanadium, yttrium, zinc and zirconium
- Alum shale, andalusite, apatite, barite, brucite, refractory clay or clinkering clay, coal, fluorspar, graphite, kyanite, magnesite, nepheline-syenite, pyrite, pyrrhotite, rock salt or other similar salt deposits, sillimanite and wollastonite
- Oil, gaseous hydrocarbons and diamonds

Other minerals, not mentioned above, belong to the landowner.

The current Mineral Act does not apply to minerals on 'public waters' (e.g. the continental shelf), and the Mineral Act administers issues related to exploration permits and exploitation concessions. An exploration permit is required if the exploration/extraction activity interferes with the rights of a landowner or tenant, etc.

In Sweden, the Environmental Code is the chief environmental statute, which aims mainly on 'promoting a sustainable development', ensuring the protection against different pollutions, environmental conservation, efficient management of land and water areas, natural resources and energy (Environmental Code, Chapter 1, Section 1). The Code provides various legal instruments to implement the objectives, for example, environmental quality standards, licensing of certain activities, EIA, protection of areas (national parks, nature reserves, etc.), remediation of contaminated areas, sanctions and enforcement tools (Michanek, 2008).

4 Development processes in disturbed ecosystems

4.1 Background

Ecosystems are complex and dynamic and involve abiotic and biotic processes interacting with the mineral substrate and the gaseous and fluid phase in the soil pore system. Ecosystems evolve over geological time-periods. An ecosystem is dynamic in its structure and function resulting from constantly evolving organisms.

The underlying regulatory processes that impact on ecosystem functionality can be explained through successional processes. Moreover, the initial stages of development affect ecosystem structure and function at later ages. A complex interaction exists between the spatial and temporal processes within the structural and functional 'components' of an ecosystem. To be able to characterize the development of an ecosystem, it is necessary to reconcile the close interaction of spatial and temporal structural assemblages that drive development.

4.1.1 Conceptual framework: disturbance

Natural communities exist in a dynamic state, as they respond to various environmental and biological conditions (Mc Cook, 1994); these may be natural or man-induced.

Events such as intensive grazing, fire and harvesting of an undisturbed ecosystem cause 'normal' disturbance. Pickett and White (1985) defined disturbance as a relatively discrete event in time and space that alters the structure of populations, communities, and ecosystems, causing change in resource availability or in the physical environment. However, on a broader scale, disturbance must be defined in the context of the normal range of environmental perturbations that an ecosystem experiences (Chapin et al., 2002). This can be explained through many

Reclamation of Mine-Impacted Land for Ecosystem Recovery, First Edition. Nimisha Tripathi,
Raj Shekhar Singh and Colin D. Hills.
© 2016 John Wiley & Sons, Ltd. Published 2016 by John Wiley & Sons, Ltd.

'types' of natural disturbance, including herbivore outbreaks, tree fall, fires, flood and volcanic eruptions. The response involves:

■ Reductions in live plant biomass
■ Altered plant community structure and composition; and
■ Sudden changes in the pool of organic matter and nutrient cycling

Disturbance is clearly not an external event that just 'happens' to an ecosystem but is an integral part of the 'environmental loading' that stimulates a response. Thus, disturbance is normal and it is extremely rare (or even impossible) for an ecosystem to remain undisturbed forever (Binelli et al., 2008). Disturbance results from both natural and anthropogenic impactors and at a moderate level the natural disturbances such as fires, grazing and hurricanes have a regulating effect on functionality (Singh et al., 1991a; Singh et al., 1991b, Singh, 1993, 1994). However, anthropogenic disturbances such as lopping, logging, mining and forest harvesting can have more profound effects. Herbivory, for example, is treated as part of the steady-state functioning of an ecosystem, whereas stand-killing insect outbreaks are treated as disturbances. However, there is a continuum in size, severity, and frequency between these two examples (Chapin et al., 2011).

Natural and human-induced disturbances affect the spatial pattern and dynamics of an ecosystem (Turner and Dale, 1998; Landres et al., 1999). According to Vogl (1980), Vitousek and White (1981), Sousa (1984), White and Pickett (1985), the following aspects of disturbance are indicating the course of ecosystem development:

■ Severity of impact
■ Shape and size of land disturbed
■ Timing (relative to season, succession and past disturbance); and
■ Spatial distribution of disturbed patches

The severity of disturbance impacts upon the survival of vegetative propagules and their contribution to the successional community (Noble and Slatyer, 1980). Similar to severity, the size and shape of disturbed land also plays a crucial role in predicting the pattern of succession. The size of the area stripped/opened by a disturbance affects the environment of a site, while its shape affects the physical environment and influences the pattern of invasion of the site (Pickett et al., 1987).

The timing of a disturbance relative to the growing season is important to succession, as it can impact more severely upon the structure of vegetation, the resources available to susceptible species and the availability of post-impact colonising species (Keever, 1979; Pickett et al., 1987). Relative to developmental status of the biological community, the timing of a disturbance can be very important if it is associated with the past history of the community (Armesto and Pickett, 1985).

Disturbance causes discontinuity in natural systems, limits resource availability and stimulates the coexistence of species (Sousa, 1979). The spatial distribution of a disturbance refers to the extent, shape and the relationship between disturbed patches of land (Seymour and Hunter, 1999). Within a given landscape or ecosystem, the spatial relationship between undisturbed and disturbed habitat is important, as re-population by impacted species depends on their seed dispersal capability and the distance between the disturbed site and the surviving populations.

Disturbance is a driving force behind succession, and this influences the rate and course of ecosystem development via impacts on the species pool, dispersal of species, and the spatial distribution of potential successional species (Pickett and Thompson, 1978; Forman and Godron, 1981).

4.2 Disturbance and ecosystem processes

Disturbance is a regulator of ecosystem processes through its impact on other key inter-related variables, including: microclimate, soil resources supply and functional organisms, and the probability of future disturbances. The ecosystem-disturbance response depends upon the severity, frequency, type, size and timing of the disturbance and their collective impact on the structure and function of the ecosystem (Heinselman, 1973).

The severity of disturbance largely determines the rate and trajectory of the succession because it regulates the supply of available resources needed to support an ecosystem. The intermediate disturbance hypothesis describes that with no or little disturbance only competitive dominant species can survive, whereas a high level of disturbance results in only fugitive species surviving (Connell, 1978; Huston, 1979; Pokhriyal et al., 2012). Therefore, an intermediate-level of disturbance supports maximum bio-diversity (Abugov, 1982), and mild disturbance promotes species turnover, colonization and a rich and diverse number of species (Whittaker, 1975). Interestingly, ecosystems are usually more resilient to frequently occurring disturbances (Thompson et al., 2009), including fire or drought as these encourage fire- and drought-adapted species that can quickly recover their biomass.

Infrequent and diverse disturbances are more likely to lead to the slow ecosystem recovery (Baskin and Baskin, 1998). For example, weedy species become adapted to the most severe conditions and produce abundant small seeds that disperse long distances, or they may remain dormant and germinate despite adverse conditions.

An important disturbance-related attribute is its impact on land-scape structure, the severity of which may impede the lateral 'flow' of both material and organisms. The frequency of disturbance impacts upon the species pool, species dispersal, and juxtaposition of succes-sional species (Pickett and Thompson, 1978; Forman and Godron, 1981). The season during which disturbance takes place is important for succession, as it can impact upon resource availability, the presence of susceptible species, and the potential for re-colonisation (Keever, 1979; Pickett et al., 1987).

4.3 Succession

Nature has its own resilience to disturbance, as natural recovery takes place to re-establish the 'pristine' ecosystem that has been impacted upon. Recovery involves succession through competition between spe-cies (Rebele, 1992), which is governed by the internal dynamics of an ecosystem and its local spatial variability. Seasonal fluctuations in eco-system processes are however, not involved.

Succession changes the composition and abundance of species in response to a particular disturbance (McCook, 1994), increasing biomass, and/or changes in life form.

Clements (1928) described succession more broadly as the growth, development and reproduction of a complex organism and assumed it to be a sequential phenomenon, whereby modification of the environ-ment supporting dominant species makes it easier for potential invad-ers to compete against the earlier occupants. This modification by plants is considered as a progressive 'reaction', leading to stabilization and supports 'relay floristics' whereby the successive dominant species occupy the space later than pioneers. If disturbance is totally absent, a unidirectional, progressive succession to a fixed climax is achieved.

Succession imparts a directional change to ecosystem structure and function. If there are no further disturbances, succession would proceed towards a climax (Clements, 1916), where a steady state/equilibrium is achieved between resource demand and resource supply. Interestingly, new disturbance usually occurs before succession reaches a climax, so an individual ecosystem stand is seldom in 'steady state' (Chapin et al., 2002).

The evolutionary process leading to succession results in a community that is governed by: the dynamics of colonization and the competition for resources from an ecological perspective the succession involves a change in community governed by these two attributes (Jenny, 1941). The succes-sional changes are most clearly delineated in primary succession, where after disturbance, the vegetation development is strongly influenced by the initial colonization events, which, in turn, depend on the environment and the availability of propagules (Egler, 1954; Connell and Slatyer, 1977; Bazzaz, 1996). In the course of succession, stability increases with the complexity of ecosystem, i.e., the number of species and the interactions between them (Pokhriyal et al., 2012).

On the basis of species density, primary and secondary succession can be differentiated. Species density declines on topsoil and ruderal soil, while increases on virgin soil during early successional stages (Rebele, 1992). Grubb (1987) distinguished secondary succession into two types: (i) internal successions that form part of the natural regeneration process in many types of vegetation and (ii) man-induced succession on old fields or clear-cut sites. Man-induced successions can be seen on mine spoil, overburden from sand and gravel extraction or on industrial waste (Borgegård, 1990).

4.3.1 Conditions for succession

Successional conditions and causes are community dynamic processes. Other changes follow the successional pattern in response to disturbance or other environmental changes. Egler (1954) suggested the term vegetation development in lieu of succession, but it could not be accepted, because the term 'vegetation' excludes many important successional processes based on sessile animals (McCook, 1994). As suggested by Pickett et al. (1987), the general conditions for succession are:

■ Availability of open site
■ Differential availability of species; and
■ Differential performance of species at the site

4.3.1.1 Initial soil formation

During the initial development of an ecosystem, time and topography influence soil and vegetational development (Wali, 1999). The accumulation of stable soil organic matter (SOM) is essential to soil development (pedogenesis), as it influences nutrient flux, and (consequently) vegetation succession. Newly deposited primary substrates contain mostly rock-derived nutrients devoid of nitrogen and organic matter. Over time, weathering of rock forms labile nutrients and minerals, which increases the biological availability of phosphorus, calcium, magnesium and potassium. Nitrogen is obtained by atmosphere, either through nitrogen 'fixation' or by atmospheric deposition (Vitousek and Farrington, 1997). Together, the physical, chemical and biological processes active in soil influence nitrogen availability (Brady and Weil, 2002; Tripathi et al., 2012).

The establishment of the initial phase of soil development is characterized by SOM development (Šourková et al., 2005). The quantity and composition of soil organic substances accumulated during initial soil formation exert an influence on the structure and physical properties of soils. Hydrophobicity may affect hydrology, structure, and the humus content of soil, and a pronounced hysteresis of the water-retention may occur (Bauters et al., 2000), resulting in preferential flow pathways (Ritsema et al., 1998), reduced infiltration capacity (Lamparter et al., 2006) and increased surface runoff and erosion (Lemmnitz et al., 2008). Hydrophobicity increases soil aggregate stability (Goebel et al., 2005) and the stabilization and accumulation of SOM (Spaccini et al., 2002).

Young degraded sites that have developed topsoil layer consisting of organic matter tend to store a considerable proportion of nutrients when added as fertilizer to the top layers (Heinsdorf, 1992). In these young soils, decomposition and mineralization of the organic matter are critical to nutrient cycling. During the early developmental phases, soils are poor in nutrients and the spatial and temporal variation in the availability of essential nutrients dominates root growth. To capture nutrients from these soils, roots grow into the close vicinity to nutrients (Robinson, 1994), and this provides an important insight into the development of soil structure (Canham et al., 1996).

4.3.2   Primary succession

Primary succession is the biological response to an extreme allogenic (periodic) disturbance (Walker, 1999), following the formation of new land surfaces consisting of rock, sand, clay substrate (where soil is totally absent). Primary succession provides insight into the loss of biodiversity, changes in climatic conditions and the influence of invasive species following disturbance.

A severe disturbance eliminate any trace of previous vegetation. Most initial colonizers (primary colonizers) have small long-distance, wind dispersable seeds and spores (e.g. algae, cyanobacteria and lichens). They form crusts and render stability to a disturbed soil (Worley, 1973). This initial colonization is followed by small-seeded wind-dispersed vascular plants (primarily woody species), however, dispersal rates depend largely on distance to the seed source (Shiro and del Moral, 1995). Late successional species tend to have heavier seeds, which arrive more slowly.

Following species colonization, the dynamic interplay between net primary productivity (NPP) and decomposition results in the availability of nutrients. During the early stages of soil development, nitrogen influences the internal cycling of other essential nutrients/elements. During mid-succession, plants and soil microbes are so efficient at accumulating nutrients, and the loss of nitrogen and other essential elements is considered negligible. In the late successional phase, inputs of nitrogen into the ecosystem may balance losses via leaching and denitrification, creating a stable nitrogen 'pool' (Vitousek and Reiners, 1975).

4.3.3   Secondary succession

Secondary succession may occur as a response to the allogenic (e.g. fire) or autogenic (e.g. herbivory) disturbance (Walker, 1999), where partial destruction of the landscape leaves the soil intact. Secondary succession differs from primary succession in that many of the initial colonizers are already present on a disturbed site (Fenner, 1987, Baskin and Baskin, 1998). Forests may have a large seedling bank that cannot develop because of the forest canopy, but grow rapidly after tree fall gaps to become the next generation of canopy dominants. Nevertheless, other colonizers, such as small-seeded, wind-dispersed species and large-seeded, animal-dispersed species also arrive from adjacent areas leading to secondary succession. Initial colonizers grow rapidly to exploit the resources made available by disturbance.

After the initial colonization, the change in species composition results from a combination of (Connell and Slatyer, 1977; Pickett et al., 1987; Walker, 1999):

- The inherent life history traits of colonizers
- Facilitation
- Competitive (inhibitory) interactions
- Herbivory; and
- Stochastic variation in the environment

Life history traits of importance include seed size and number available, potential growth rate, maximum size and longevity, and the ability to finally survive. Most early secondary successional species arrive soon after a disturbance, grow quickly, are relatively short-lived and have a low maximum longevity, compared to late successional species (Noble and Slatyer, 1980). Even in absence of species interactions during succession, a life history pattern alone would cause shift in dominance from early to late successional species because of differences in arrival rate, size and longevity.

Under severely impacted environments, facilitation becomes particularly important and involves processes in which early successional species make the environment more favorable for the growth of later successional species. By way of example, during succession, species ameliorate soil conditions through the addition of organic matter and fixing of nitrogen, creating favorable conditions for the establishment of later successional species (Callaway, 1995; Brooker and Callaghan, 1998).

There are four main phases identified in secondary succession, namely reorganization, aggradation, transition and steady state (climax). The 'reorganizational phase' is the period immediately after the establishment of pioneer species, when the resources (light, moisture and nutrients) are abundant and species competition is low. However, because of fewer leaves per unit area, loss of water from leaves and uptake of nutrients by plants is low, and the runoff of water is high, which cumulatively lead to heavy nutrient loss from soil. During the 'aggradation phase', the accumulation of biomass by plants is rapid and detritus forms on the ground. The transition phase is characterized by an initiation of tree mortality, caused by increased competition among the pioneer tree species, accumulation of wood biomass, and the establishment of shade tolerant understory species. Finally, large accumulations of living biomass and debris (wood biomass, litter mass tree death) occur followed by slow growth of tree species. At this phase the ecosystem attains highest structural diversity and stability (Bormann and Likens, 1979).

Immediately after disturbance, NPP is low because of paucity of plant biomass. However, the initial C pool and its availability is higher than during primary succession because of the presence of initial stock of SOM and nutrients, augmented by the labile carbon added due to the death of leaves and roots.

Herbaceous species play a very crucial role in increasing NPP, as they return most of their biomass to the soil. With an increase in perennial plants, particularly woody species, biomass and NPP increase rapidly however, the changes in biomass in NPP remain similar (to the primary succession), as the controlling factors are similar except the biomass decomposition rate that is rapid at this time.

Many disturbances, such as tree fall and insect outbreak, transfer large amounts of labile carbon to soils and create an environment that is favourable for decomposition (Chapin et al., 2011). Due to removal of vegetation, the soil surface is exposed to more sunlight, and transpirational loss is low, which leads to an increase in soil temperature and moisture, and decomposition. The large quantity of litter formed by early secondary successional plants also promotes decomposition.

The decomposition of biomass and the accumulation of carbon depend on initial carbon stock, the availability of resources, the severity of the disturbance and the successional pathway (Lorenz and Lal, 2010). During early secondary succession, NPP is low and carbon loss is high. As the succession progresses towards mid-succession, NPP overtakes decomposition and ecosystem starts accumulating carbon, which during late succession continues at a slow rate.

After disturbance, the dynamics of nutrient loss is highly variable and dependent upon the capacity of the regenerating vegetation to utilize/assimilate nutrients (Chapin et al., 2011). During secondary succession the availability of nitrogen is higher and a pulse of nutrients may result from an increase in litter and the mineralization of organic matter. Plant growth is, however, not generally strongly nutrient-limited early during secondary succession as there is usually adequate nitrogen to support high rates of photosynthesis and growth (Scatena et al., 1996).

4.4   Ecosystem development in mine spoils

Vegetation strongly influences initial soil development. The basic conceptual framework of soil formation and terrestrial ecosystem development is mainly based on changes in soil properties over time (Dokuchaev, 1893; Jenny, 1941; Yaalon, 1975). Initial soil development and plant succession are interdependent and control each other (Odum, 1971; Jenny, 1980). In addition, the rate and direction of ecosystem development are controlled by both exo- and endogenic factors, which help soil system to achieve structural properties and gain new functional stages (Schaetzl et al., 1994; Bockheim et al., 2005).

After the dumping of mine spoil, the substrate is initially free of pedogenic organic carbon accumulated by biomass. Mine spoil surfaces are often dry, with high surface temperature, dark in colour and lack water. Generally, mine spoils have higher bulk density, poor soil structure, low porosity, low soil moisture, and water holding capacity (Tripathi and Singh, 2008). Spoil is often very compact due to movement of heavy machinery, impoverished carbon and nutrients and low cation exchange capacity. These characteristics make them unfavourable for vegetation development.

A comparative study in spoil with undisturbed soil of North Dakota with respect to species composition and soil characteristics showed that the spoils had higher pH, electrical conductivity, exchangeable magnesium and sodium, total phosphorus, sulphur and more silt and clay (Wali and Freeman, 1973). Generally lower clay content in spoils compared to undisturbed soils indicates difference in parent material. The soil development processes in mine spoils include accumulation of organic matter and formation of associated structure; redistribution of carbonates and soluble salts and finer-textured horizons due to clay formation. The most tertiary mine spoils, the presence of geogenic organic carbon (e.g. lignite or fossil carbon) also helps in soil redevelopment through the accumulation of organic matter and nitrogen via microbial decomposition (Katzur and Hanschke, 1990; Laves et al., 1993; Waschkies and Hüttl, 1999) and increased soil water holding capacity (Thum et al., 1992; Embacher, 2000).

The goal of ecosystem development via mine land reclamation is restoration of a disturbed land to its pre-mining or 'higher or better uses' by establishing a diverse, effective, permanent vegetative cover of native species which reduces the reliance on natural succession. This facilitates productivity and stability against erosion. With respect to the development of ecosystems on mine sites, the major focus should be on pedogenesis, because due to open cast mining, soil is the most affected component.

From an ecological point of view, ecosystem development in post-mining areas offers the rare opportunity to examine the development of ecosystems starting at 'point zero' (Bradshaw, 1983). In addition, post-mining landscapes are known for being unique habitats frequently inhabited by rare species (Abresch et al., 2000). Thus, acceleration of natural successional processes to increase biological productivity, soil fertility and biotic control over biogeochemical fluxes within the disturbed mine spoil system could be a tool for their recovery and ecosystem development.

Initial soil development may vary widely, depending on the textural and mineral composition and the spatial arrangement of the parent material. However, during the development of a soil, key structural properties are acquired resulting from mineral weathering, the development of soil horizons, and the formation of stable secondary minerals (Schaetzl et al., 1994; Bockheim et al., 2005).

4.5   Options in restoration

Restoration ecology can be viewed as an attempt to speed successional processes to reach a desired climax community. Restoration managers can explore different mechanisms of succession to rapidly achieve climax conditions by greatly increasing seed availability, reducing competition by early successional species, and amending soil to match late succession conditions. While not always successful, restoration efforts are often seen as an acid test for our understanding of succession (Young et al., 2005).

A damaged ecosystem has its structural and functional components impaired. Although it is impossible to regain an original (pre-disturbance) state, managed re-vegetation can be used as a 'close to original' restoration strategy. Since natural recovery is a long process, re-vegetation process can bypass natural succession. Thus, during restoration, fast growing trees and shrubs can be planted to shorten the reorganizational phase and hasten the aggradation phase (Binelli et al., 2008).

The process of restoration is progressive. Bradshaw (1996) suggested that ecosystem development should be an unrestricted upward path. In restoring a damaged ecosystem, it is quite difficult to restore the original structure; however, the functions are restored completely. Considering these facts, true restoration seems to be unrealistic.

4.5.1   Re-vegetation objectives

Disturbances facilitate the establishment of additional species to expand the vegetation composition. Vegetation succession studies depict that the conditions just after disturbance (i.e. at 'point zero') are important determinants for further vegetation development. According to the initial floristic concept, the first established species tend to occupy all suitable growing sites and sustain the conditions at very beginning (Schaff et al., 2011). Consequently, they regulate the species and vegetational diversity over longer times. With progressing succession and the improving soil conditions, successive species become dominant. Changes that occur in major structural and functional characteristics of a developing ecosystem are listed in Table 4.1.

Table 4.1 A tabular model of ecological succession: trends to be expected in the development of ecosystems (Odum, 1969).

Ecosystem attributes	Developmental stages	Mature stages
Community energetics		
Gross production/community respiration (P/R ratio)	Greater or less than 1	Approaches 1
Gross production/standing crop biomass (P/B ratio)	High	Low
Biomass supported/unit energy flow (B/E ratio)	Low	High
Net community production (yield)	High	Low
Food chains	Linear, predominating grazing	Web-like, predominantly detritus
Community structure		
Total organic matter	Small	Large
Inorganic nutrients	Extrabiotic	Intrabiotic
Species diversity: variety component	Low	High
Species diversity: equitability component	Low	High
Biochemical diversity	Low	High
Stratification and spatial heterogeneity (pattern diversity)	Poorly organized	Well organized
Life history		
Niche specialization	Broad	Narrow
Size of organism	Small	Large
Life cycles	Short, simple	Long, complex
Nutrient cycling		
Mineral cycles	Open	Closed
Nutrient exchange rate, between organisms and environment	Rapid	Slow
Role of detritus in nutrient regeneration	Unimportant	Important
Selection pressure		
Growth form	For rapid growth (*r-selection*)	For feedback control (*k-selection*)
Production	Quantity	Quality
Overall homeostasis		
Internal symbiosis	Undeveloped	Developed
Nutrient conservation	Poor	Good
Stability (resistance to external perturbations)	Poor	Good
Entropy	High	Low
Information	Low	High

The primary objectives of re-vegetation programmes are to:

- Provide an erosion-resistant plant cover on overburden dump slopes
- Focus on utilization of native woody-stemmed reclamation species common to the region
- Strive to establish a diverse range of plant species to recreate the level of biodiversity similar to the pre-disturbed site; and
- Establish a viable plant community capable of developing into a self-sustaining cover of species suitable for commercial forest, wildlife habitat, traditional land uses and with possibilities for recreation and other end uses

4.5.2 Implications for management

It may not be possible to achieve an original soil profile in less than 5000 years, although the biological functions of a soil can probably be restored in less than 10 yr (Bradshaw, 1996). Therefore, rehabilitation and replacement may be appropriate, and may provide an 'end point' for restoration of disturbed land. Replacement may allow restoration of a component (e.g., productivity, to a higher level than that existed previously), or an attribute such as greater biodiversity.

To achieve restoration or rehabilitation, it is important to consider the following hypotheses, as suggested by Aronson et al. (1993):

- Beyond one or more thresholds of irreversibility, ecosystem degradation is irreversible without structural intervention combined with revised management techniques
- The more of these thresholds that are passed, the more time and energy is required for restoration or rehabilitation
- Without massive intervention, restoration will proceed only as far as the next highest threshold in the process of vegetation change or succession
- Beta diversity (the extent of species replacement or biotic change along environmental gradients) and life form ranges decline with ecosystem degradation, while alpha diversity (species richness of a place) temporarily increases
- The loss of keystone species speeds degradation more than the loss of other species, and tends to result in irreversible change
- The re-introduction of keystone species should accelerate rehabilitation of an ecosystem by the establishment of additional native species
- Water and nitrogen use efficiency and nutrient cycling times decrease with ecosystem degradation
- Diversity of soil biota and their compatibility with extant higher plants decrease with ecosystem retrogression

- Between a floor and a ceiling of a given phase of retrogression, resistance increases but resilience decreases; and
- The rate of recovery in restoration or rehabilitation pathways is inversely related to the structural and functional complexity of the ecosystem of reference

Restoration requires the consideration of historical, social, cultural, political and aesthetic aspects (Higgs, 1997), and is executed by the historical ecological knowledge (White and Walker, 1997; Egan and Howell, 2001).

Ecological restoration focuses on restoration of entire ecosystem or community, or on the rescue or reintroduction of certain target species (Andel and Grootjans, 2006). In order to achieve ecosystem development through restoration, indicators can be used to explore the possibilities of success. The organisms (e.g., terrestrial plants) are employed as indicators in exploring new situations or evaluating large areas. In the Western United States, for example, plants are used as indicators of water and soil conditions (especially as they affect razing and agricultural potentials). The use of vertebrate animals, as well as plants, has been reported to indicate temperature zones. For example, steno species (unable to tolerate a wide range of temperature) are much better indicators for temperature than eury species (organisms that can tolerate a wide range of temperature). Further, large species usually make better indicator than small species (Odum, 1959).

Numerical relationships exist between species, populations, and whole communities providing more reliable indicators than single species, since a better integration of conditions is reflected by the whole than by the part. This has been particularly well established with respect to the biological indicators of various sorts of pollution. Indicator lists of varying length seek to capture the different – economic, environmental, social and institutional – dimensions of sustainable development. Indicators provide early warning of non-sustainable economic activity and environmental deterioration. They can also support policy formulation and evaluation, and the setting of targets or benchmarks against which progress and/or failure can be assessed (Bartelmus, 1997).

Indicators may reflect different ecosystem conditions at a micro-scale. Additionally, indicator species are essential for ecosystem functionality, and for the quality control of the environmental conditions (Andel and Grootjans, 2006). As an indicator of soil development, microbiological indices are more appropriate options to measure damage and degree of restoration. Microbial biomass plays a crucial role in effective element turnover in soil and import and export of nutrients for plants (Singh et al., 1989). There exists a close relation between ecosystem development, plant community composition and microbial community structure and function (Banning et al., 2011). Microbial diversity and their structural and functional characteristics (e.g., biomass, nutrient transformation and turnover potential) within the system can be used as efficient indices to measure ecosystem damage and repair. These microbial parameters depict the qualitative and quantitative successional changes in available organic matter to the soil microbes (Harris et al., 1996).

During the initial phase of ecosystem development, the microbial community structure and functions develop showing an increase in biomass over time (Sigler et al., 2002; Tscherko et al., 2004). Ammonia and nitrate in soil are limited and inhibit plant growth. Microbes play a crucial role in rendering an efficient mobilization and transport of bio-available forms of N to plants (Xu et al., 2012). Albeit the diverse links between plant and microbial communities are still missing, plants are the drivers for the evolution of microbial community structure in soil (Bardgett and Walker, 2004).

Biological components reflect the dynamism and constantly respond to both external and internal changes. Spontaneous, natural colonization of mine sites by flora, fauna and microorganisms can be observed within the first year after rehabilitation. This colonization also includes symbiotic microorganisms such as mycorrhizal fungi or N-fixing bacteria (Kolk and Bungart, 2000). A number of species that invade mine sites are extremely rare at non-mined sites (Golldack et al., 2000).

The re-vegetation of dumped mine spoil and overburden is not an operation to be considered only at or just before, mine closure. Rather, it should be integrated into an effective environmental management system extending from exploration and mine construction to its operation and closure. The recovery of mine soil quality and ecosystem function of disturbed site depends on re-establishing vegetation and biodiversity that can thrive and sustain itself.

5 Benefits of reclamation

5.1 Background

The restoration of mining-impacted land is an integral part of the management of a mining site. Following mineral extraction the land must be able to support a sustainable ecosystem. The use of a site-specific environmental management plan will minimize the impact on the environment and conserve biodiversity for the benefit of society at large.

Reclamation brings derelict, degraded and damaged land back or close to its original condition through the restoration of functional ecological structures, through the:

- Formation of a suitable soil texture
- Maintenance of soil fertility; and
- Sustainable microbial populations

The re-introduction of a functional ecosystem is reliant on a viable soil-microbe interactions and sustained nutrient cycling. To return land to a condition,which supports a self-sustaining ecosystem, the following are important (Jordan et al., 1990):

- Stabilization of land surfaces
- Pollution control
- Aesthetic improvement
- General amenity; and
- Plant productivity

Reclamation of Mine-Impacted Land for Ecosystem Recovery, First Edition. Nimisha Tripathi, Raj Shekhar Singh and Colin D. Hills.

More ambitiously, the restoration of the ecological components that were originally present will secure:

■ Biodiversity (species composition); and
■ Ecosystem function

The re-introduction of a viable soil for re-vegetation and succession relies upon the ability to support a plant community supported by microbial processes and a stable nutrient cycle. A sustained vegetative growth improves soil texture, particularly in compacted or loosely consolidated soil as it imparts aeration and improves water infiltration limiting surface water run-off and slope erosion. The 'ideal' range of soil bulk densities for restoration purposes can be found in Table 5.1.

A stable and self-sustaining vegetative cover has two basic ecosystem-related components, termed 'structure' and 'function': The structure (of an ecosystem) can be measured as a mono-specific grass, or be a tropical rainforest, whereas, the function relates to the processes supporting soil productivity and nutrient cycling; these can be 'low' or 'high'.

Bradshaw (1987a) identified the importance of a stable and self-sustaining vegetative cover, as it is a regulator of the physical properties and biological diversity of a disturbed site.

■ A very low pH (in the mine spoil)
■ An extreme phosphorus deficiency (due to high pH); and/or
■ A high calcium concentration

In a disturbed ecosystem, three principal mechanisms determine the process of succession. These are: facilitation, tolerance and inhibition (Connell and Slatyer, 1977). Facilitation occurs when early colonizing species create conditions that facilitate the establishment of other species that cannot occur in their absence. Both tolerance and inhibition influence the establishment of later-colonizing species.

In natural ecosystem development, plant species establish slowly aided by the physical and chemical changes occuring during primary succession. They are dependent on the availability of native propagules and can also exploit favourable conditions for colonization over a temporal scale. However, re-vegetation strategies involve the use of established forestry techniques.

Table 5.1 Ideal bulk density of different textural classes.

Soil texture	Ideal bulk density for plant growth (g cm^{-3})	Bulk density that restricts root growth (g cm^{-3})
Sandy	<1.60	>1.80
Silty	<1.40	>1.65
Clayey	<1.10	>1.47

5.2 Establishment of ecological succession

The re-vegetation of mine spoil by tree cover can be used to stabilize an ecosystem over the long-term. Tree cover can ameliorate soil quality, and improve the commercial and aesthetic value of disturbed soil. Plant species can exert a catalytic effect by changing underlying microclimatic conditions, as they impact on the physical and chemical properties of soil. The complexity of vegetative cover and the development of litter and humus accelerate the development of biota at a degraded site. Plantations accumulate fine particles and reverse degradation process through root proliferations.

However, adverse impacts result from excessive acidity, poor nutrient supply, increased toxic heavy metals and the poor soil structure. Poor conditions will inhibit plant growth, however, the use of non-native plant species for the restoration of less severely degraded land, is an acceptable option to provide a temporary successional role prior to the colonization and eventual dominance by native flora. Figure 5.1 illustrates the invasion of coal mine spoil by non-native plant species.

A plant succession, adapted to the conditions of mine spoil will, build SOM and provide nutrients to support a self-sustaining cover; thereby accelerating the natural recovery process (Figure 5.2). The shedding of plant leaves and roots builds SOM and microbial colonies fix nitrogen to regulate soil chemistry.

Soil fauna/microbial activities enhance soil structure by changing the soil-water holding capacity and air content, whereas the succession

Figure 5.1 Exotic species growing on tipped overburden.

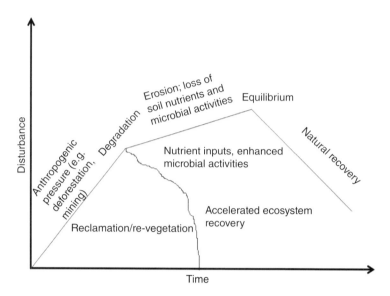

Figure 5.2 Natural and accelerated recovery of a damaged ecosystem (Sheoran et al., 2010).

of plants tends to reduce soil surface temperature and raise soil moisture. Cumulatively, these 'biotically' induced changes imparted by soil microbial and animal communities lead to the establishment of a diverse ecosystem (Sheoran et al., 2010).

5.3   Recovery of damaged ecosystems

The disruption of natural ecosystems by mining and their decline in productivity drive the restoration of soil functionality. However, as spoil-dump height is linked to slope failure, it is necessary to manage slope stability to minimise the occurrence of damage/disruption. This can include engineering the mine-spoil profile and the use of vegetative cover. Grass can be effective against sheet erosion and small rill erosion, whereas deep-rooted plants, trees and shrubs can provide structural reinforcement to mine spoil.

As a disturbed ecosystem may take a long time to reach an equilibrium state (Sopper, 1992), recovery is faster in a re-vegetated mine spoil (Tripathi et al., 2012), particularly when diverse species are planted, as they enhance soil fertility by increasing soil-microbe-plant interactions. Plants act as natural bio-engineers by increasing soil organic matter, improving soil pH, bulk density and moisture content, and by bringing mineral nutrients to the soil surface in an available form.

Deep-rooted trees grow slowly and help mineralise nutrients on the soil surface. Tree roots act as scavengers of less-available nutrients and accumulate soil surface organic matter for microbial breakdown. Soil microbes physically connect mineral particles and organic materials together and this helps stabilize soil and enhances soil pedogenic processes, which over enough time may meet the soils pre-mining morphology.

In general, however, different organic amendments (e.g., sewage sludge, domestic refuses) or inorganic (fertilizers) are necessary to establish plants on mine spoil. Acidic mine soils can be effectively neutralized by alkaline amendments, including cement kiln dust (CaO) or agricultural lime (limestone; CaCO3), but at rates that account for oxidation of pyrite oxidation so as to maintain a neutral soil pH. Gravellia robusta, Dendrocalamus strictus, Shorea robusta and Dalbergia sissoo can be planted on acidic dump soil (pH 3.6–3.9), but organic amendments such as woodchip, composted green waste or manure and biosolids can be used to increase soil pH; improve soil structure/water holding capacity, and cation exchange capacity; provide slow-release fertilizers and serve as effective microbial inocula (Tordoff et al., 2000).

Gypsum ($CaSO_4.H_2O$) has traditionally been used to improve sodic soil conditions for plant growth. Gypsum is normally incorporated into soil at about 5–10 te ha^{-1}. It is an effective replacement of sodium (by calcium) in clay mineral exchange surfaces with the consequence of improving soil structure, reducing surface crusting and increasing water infiltration. The reduction in pH by gypsum in sodic soils where pH is >8.5 and exchangeable sodium is >6% is particularly effective (Ghose, 2005).

The growth of plants on re-vegetated mine spoil depends on the metal tolerance of the colonizing species in addition to the presence of available nutrients, especially N and P.

When soil moisture, nutrient supply and aeration are not limiting factors, high germination rates can lead to rapid re-vegetation. The addition of woodchip, forest litter or organic municipal waste to bare spoil is second only to the use of topsoil. Amendments with wood residue increase the effect of fertilizers (N, P, K, etc.), Whilst gypsum increases the level of soluble salts. The nitrogen needed by a developing plant/soil community comes from the mineralization of organically combined N, and this can be promoted by incorporation of a significant component of legumes in the re-colonising plant community.

However, it should be noted that most mine soil/spoil does not contain native populations of the essential N_2-fixing *Rhizobium* bacteria that enable legumes to capture atmospheric N. Therefore, as N is primarily combined in organic matter in soils, the addition of organic amendments will greatly enhance total N in soil over time.

As envisaged by Bradshaw (1987a), the toxicity of heavy metals and deficiency of major nutrients are often limiting factors for establishing plants on mine tailings. Therefore, the success of restoration schemes involves the introduction of species that are tolerant to increased metal exposure and deficient nitrogen. Thus, to achieve long-term stability, metal tolerant species are commonly used with mine tailings, with herbaceous legumes applied as the pioneer species. As mentioned, leguminous plants alleviate nitrogen deficiency in mining soil, whereas mycorrhiza improves the supply of phosphorus in damaged soil. According to Tripathi and Singh (2008), it should be possible to produce biomass at the levels found in a pristine forest ecosystem after a period of 30 years, after re-vegetation has been carried out.

5.3.1   Biological macro-aggregate formation

Seedlings develop a dense root community and these are accompanied by the formation of soil aggregates which enhance erosion resistance, and maintain C and N. Soil aggregates physically protect SOM by:

- Forming a physical barrier between microorganisms/microbial enzymes and their substrates
- Controlling food web interactions; and
- Influencing microbial turnover

The processes involved in aggregate formation and stabilization in temperate versus tropical soils show there is a close relationship between soil biota and SOM dynamics.

The formation of unstable macroaggregates: There are several biological processes responsible for the formation of unstable 'biological' macro-aggregates in both temperate and tropical soils (Six et al., 2002; Fig. 5.3):

- Fresh plant- and root-derived residues form the nucleation sites for the growth of fungi and bacteria (Puget et al., 1996). Fungal hyphae initiate macro-aggregate formation by enmeshing fine particles into macro-aggregates (Tisdall and Oades, 1982). Microbial (bacterial and fungal) exudates, produced as a result of decomposition of fresh residue, form binding agents that further stabilize macro-aggregates (A1, A),
- Biological macro-aggregates also form around actively growing roots in both temperate and tropical soils. Similar to fungal hyphae, roots can provide the mechanical framework for initial formation of macro-aggregates by enmeshing particles and production of cementing agents (root exudates), which stimulates microbial activity (A0, B–A1, B) (Jastrow et al., 1998).
- Another mechanism of biological macro-aggregate formation in temperate and tropical soils is through the activity of soil fauna, that is, earthworms (through their organic-rich casts) (Marinissen and Dexter, 1990), ants and termites (Brauman et al., 2000) (A1, C). However, casts

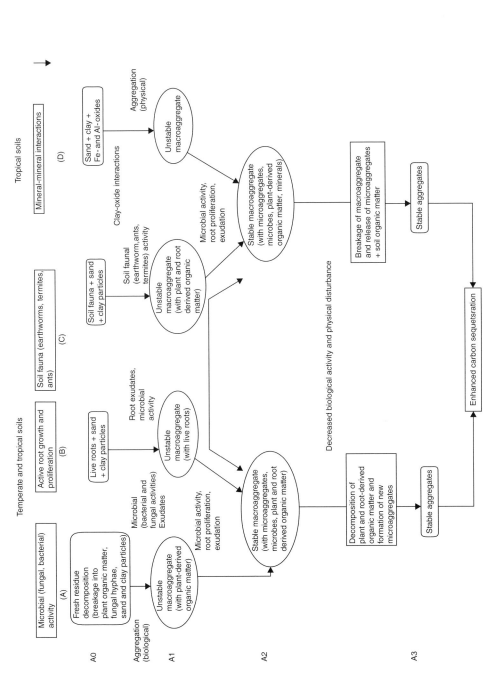

Figure 5.3 Aggregate formation and degradation mechanisms in temperate and tropical soils (Six et al., 2002). Fungal, bacterial and earthworm activity augmented by active root growth are the biological aggregate-forming mechanisms in both temperate and tropical soils, whereas mineral-mineral interactions are the physicochemical mechanisms forming aggregates in tropical soils.

are not stable when they are freshly formed and wet (Marinissen and Dexter, 1990). When earthworms ingest soil and particulate organic matter, large amounts of watery mucus are added and the soil undergoes a thorough kneading. This moulding of soil breaks the bonds between soil particles, thereby reducing stability (Griffiths and Jones, 1965; Utomo and Dexter, 1981). However, the close contact between organic matter, mucus and soil particles in casts can lead to highly stable casts upon drying (Marinissen and Dexter, 1990).

The formation of stable aggregates, or those that have the capacity to withstand air drying and quick submersion in water (a process called 'slaking') are formed (A1 to A2):

- Under wet conditions, over a period of time, the microbial activity facilitates binding of macroaggregates with an addition of polysaccharides and other organic materials.
- During dry-wet cycles the primary particles are rendered closer arrangements, with a consequent increased aggregate stability (Kemper and Rosenau (1984).
- The actively growing roots accelerate the stabilization process by producing exudates. Acting as a cementing agent, the exudates adsorb the soil inorganic fractions and also stimulate production of microbial binding agents. Finally, the roots induce compaction and remove water via transpiration to stabilize the aggregates.

Macroaggregate can be stabilised when fresh plant residues are incorporated in their structure, as the breakdown of organic matter results in its encapsulation by soil minerals and microbial products. These form as new microaggregates (53–250 μm) within the macroaggregate structure and have a stabilising effect. The formation of microaggregates in this way is crucial for the long-term immobilisation of carbon, because the microaggregate compsoites have a greater capacity to protect C from decomposition, in comparison with a macroaggregate itself (Skjemstad et al., 1990). As a soil further develops aggregates get turned over, or recycled (A2 to A3) and microaggregates are released and microbially processed as SOM.

In the tropics, plantations with either native or non-native species have potential to 'foster ecosystems' by modulating extreme microclimates improving soil nutrient availability and the provision of habitats for seed-dispersing animals.

5.3.2   Enhancement of soil fertility

The enhancement of soil fertility is one of the major objectives when restoring disturbed soil. The addition of organic materials such as plant litter and/or root growth in soil stimulates microbial activity, promotes N transformation and nutrient cycling. When the clay content of soil is greater than 40% w/w, the soil has reduced permeability, a lower infiltration rate and impaired structure. When the sand content of a soil is >70% w/w, the soil will not retain sufficient water to allow plants to grow and develop (Dollhopf and Baumm, 1981; Singh and Singh, 1991).

The long-term productivity of a soil is dependent upon several major factors: (i) accumulation of SOM and N, (ii) maintaining N-fixing legumes in the soil and (iii) establishment of an organic P pool and avoidance of P-fixation. Additions of organic matter improve soil structure, reduce erosion and increases infiltration. Sawdust and bark mulch help in increasing the initial mine SOM contents but are generally low in N. Sawdust and sewage sludge provide a microbial inoculum and are the most effective short-term fertilizers and sources of long-term slow-release nitrogen (Munshower, 1994; Sydnor and Redente, 2002).

The biological *protection* of soil can be attained via the involvement of microbes and earthworms by initiating the formation of micro-aggregates (within large macro-aggregates) leading to long-term stabilization of SOM against microbial decay (Pulleman and Marinissen, 2004; Bossuyt et al., 2005).

The aggregation of soil strongly influences its hydraulic properties, the amount of nutrients that are available and the pathway for the long-term stabilization of organic carbon. The stability of macro aggregates influences soil macro-porosity and its ability to transmit water. The ability of microbes and soil biota to promote aggregation through the excretion of binding agents and fecal pellets (Lynch and Bragg, 1985) is augmented by root exudates, which flocculate colloids further binding soil into or stabilised aggregates (Glinski and Lipiec, 1990). Consequently, inoculation with mycorrhizae can enhance plant-available nutrients, particularly phosphate-P.

5.3.3   Establishing a nutrient supply

Plant nutrient 'pools' exist in different forms within soil (Fig. 5.4) including: (a) soluble to weakly-bound forms, that are readily available or are in rapid equilibrium with soluble pools, to (b) strongly-bound and precipitated forms, that are very insoluble and become available only over long time periods, respectively. The availability of nutrients is influenced by soil pH and how weathered it is, but in general, mine spoil is deficient in nitrogen and phosphorus.

As the availability of nitrogen is a major limiting factor for the re-vegetation of spoil, it should be noted that others may also be present that maybe physical, chemical or biological in nature. According to Singh and Singh (2006), there is a relationship between soil N and the quantity of the N deposited by litter fall. This indicates that the fertility

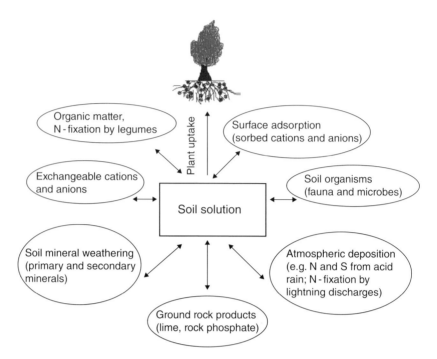

Figure 5.4 Generalized nutrient pools in soil (Bierman and Rosen, 2005).

is indeed a function of the N returned by the vegetation to the soil. Interestingly, nitrogen does not occur as a soil mineral, but is supplied by the decomposition of the organic matter and species which allocate a greater proportion of biomass to foliage producing N-rich litter and accelerate the nutrient supply and soil fertility in revegetated mine spoil (Raghubanshi et al., 1990; Singh and Singh, 2006).

As mentioned, fresh spoil is devoid of N and organic matter and is often low in water-soluble phosphorus. As spoil is weathered, pH may decrease making P available. However, iron oxides may also be formed and may 'fix' P, thereby limiting its availability (Liptzin and Silver, 2009).

To maintain a stable biological 'system', which is productive over time, the rapid accumulation of organic matter, nitrogen and organic phosphorus, and N-fixing legumes are crucial. A stable N-cycle is particularly important to tree growth, and the introduction of topsoil may increase available nitrogen and other nutrients contained in new organic matter and promoting the growth of microorganisms capable of mineralizing nitrogen (Woodmasee et al., 1978).

The nitrogen mineralization potential of a soil is an indicator of its fertility/productivity (Tripathi et al., 2008) as N is an important growth medium to plants. It is reported that re-vegetation can significantly enhance annual net N-mineralization (by 2–3 times) and annual net nitrification (by 3–5 times) compared to adjacent virgin (not-mined) land (Kaye and Hart, 1998).

The leguminous tree *Leucaena leucocephala* can 'fix' about $100 \, kg \, ha^{-1} \, year^{-1}$ nitrogen (Wild, 1987). *Acacia* spp. and *Albizia lebbeck* are also promising N-fixing species (Sharma and Sunderraj, 2005) due to the high nitrogenous activity of their root nodules.

Nitrogen-fixing plant species have a dramatic effect on soil fertility as they readily produce decomposable nutrient-rich litter and turnover fine roots and nodules. The mineralised N from this litter is also available to companion species enhancing the development of a self-sustaining ecosystem. Native leguminous species improve soil fertility more than native non-leguminous species, with native legumes being the more efficient than exotic legumes in the shorter-term (Singh et al., 2002).

The establishment of biological systems that cycle the soil nutrients is critical to the development of stable ecosystems, with new surface soils planted with N-fixing trees and legumes being the preferred approach. The fertilisation of N deficient soils, in the range from 50 to $75 \, kg \, ha^{-1}$, will allow trees to respond without enhancing the growth of competing vegetation. The establishment of a long-term phosphorus supply should be greater than $100 \, kg \, ha^{-1}$, and this can be easily provided by application of farmyard manure (contains nitrogen, phosphorus and potassium).

5.3.4   Remediation of heavy metals

Mine spoil generally has elevated levels of phyto-toxic elements, which can be more important to address than the availability of nutrients, as metal toxicity can lead to a sparsity of vegetation over long time periods.

Although a covering of organic matter might effectively bind some metals in the short term, it will normally decline after a few years (as the organic matter disappears) and toxicity returns. A simple protecting mechanism might involve using a cover system involving inert material acting as a barrier to the upward movement of metals and the downward growth of roots (Smith et al., 1985; Williamson and Johnson, 1991).

Contaminated mine soil can be remediated by the application of physical, chemical or biological techniques as:

(a) *Ex situ* treatments, which require removal of a contaminated soil for treatment on- or off-site
(b) *In situ* treatments, which remediate a contaminated soil without its excavation

In situ techniques are favoured due to their lower cost and reduced environmental impact. On-site management of heavy metal-contaminated soils can be achieved either by dilution to safe levels by using clean soil (Musgrove, 1991) although it should be noted that this practice is not now widely permissible, or by stripping and stockpiling clean topsoil, and redistributing it over the affected area. Similarly, if environmental regulations permit, the vertical mixing of heavily con-taminated soil with less contaminated subsoils through deep ploughing can also be used to dilute the heavy metal contents (Thompson-Eagle and Frankenburger, 1992).

The immobilization of inorganic contaminants is also possible (Mench et al., 1994) by complexing the contaminants (Wills, 1988) or by increasing the soil pH by liming. The solubility of priority metals, such as Cd, Cu, Zn and Ni, can be reduced by the formation of insoluble hydroxides (Adriano, 1986). Soil washing/metal extraction is an alternative to off-site burial method (Elliott et al., 1989; Tuin and Tels, 1991). A number of cement-based systems to stabilise/solidify or other physico-chemical methods are also available for use as in-situ and ex-situ application.

SOM can adsorb heavy metals, increase water infiltration rates and reduce erosion. Lime addition can decrease heavy metal mobility in soils and their phyto-accumulation in the plant (as pH increases). Vegetation cover inhibits the formation of acidic biological complexes, through the uptake of oxygen/moisture by root systems, reducing the activity of acid-generating bacteria (Ledin and Pedersen, 1996).

Re-vegetation moderates pH and nutrient levels and lowers the levels of available toxic elements, through binding in root systems and foliage. The reductions in metals, bioavailability protects the surrounding envi-ronment (Park et al., 2012).

Re-vegetation may be appropriate for certain metals, petroleum-based products, chlorinated hydrocarbons and polyaromatic hydro-carbon (PAH) compounds, but must consider:

- Quantification of soil pH, nutrients and organic matter content of the mine spoil and wasteland, moisture conditions.
- Characterization of hazardous compounds and their respective concentrations.
- Selection of suitable plant species.
- Root systems must not form pathways for water infiltration and metal transport.
- The addition of fertilizers should not lead to changes in pH that might liberate or leach metals or other hazardous compounds; and
- Removal of tolerant plants accumulating hazardous compounds.

Plants have evolved various effective mechanisms to tolerate high con-centration of metals in soil, including their transport by breakdown in roots (belowground) or in woody tissue and foliage (aboveground). The plants endemic to metal-contaminated soils tend to possess metal tolerance as an intrinsic property and they are adapted for elevated levels of heavy metals (e.g. Zn, Cu and Ni) such as those present in mine spoils.

Some plants have a limited potential for accumulating metals and their tolerance is achieved by preventing their uptake into root cells. Accumulator plants readily imbibe metals into their roots, because they possess specific mechanisms that tolerate their presence in cells.

Another group of plants, termed indicators, have little control over metal uptake and accumulation and therefore indicate the metals concentration in rhizospheric soil and this allows them to be used for prospecting new ore bodies (Raskin et al., 1994).

The maximum amount of metal ions associated with roots is limited by absorption into their cells. A significant ion fraction is physically adsorbed at the extracellular negatively charged sites (COO$^-$) of the root cell walls. The cell wall-bound fraction cannot be translocated to the shoots and, therefore, cannot be removed by harvested. As such, plants with significant metals accumulated in their roots are not suitable for phytoextraction. As metals can be complexed and sequestered in cellular structures (e.g. vacuole) they simply become unavailable to plant shoots (Lasat et al., 1998).

The uptake of metals into plant root cells is important for phytoextraction; however, they must be transported to the shoots by the movement of sap. This is primarily controlled by two processes: root pressure and leaf transpiration (is termed translocation), followed by re-absorption from the sap into leaf cells. A schematic representation of metal transport processes is shown in Figure 5.5.

Re-vegetated plant species vary widely in their tolerance to acidic environments and the presence of toxic metals. Plants are harmed directly by acidity at pH <3.0 (Arnon and Johnson, 1942), partially as a result of the increased availability of metals including aluminium and manganese.

Vetiver grass (V. zizanioides) is one of the most commonly used aromatic grass species for reclamation, as it can survive very harsh conditions including high temperatures. It has a massive finely structured and deep root system capable of reaching >3 m in the first year of growth. Due to its unique morphological and physiological characteristics, it is highly effective for preventing erosion under extreme conditions e.g. prolonged drought, flood, extreme temperatures (−10 to 48 °C), and over wide range of soil acidity and alkalinity (pH 3 to 10.5) (Greenfield, 1989). Vetiver is also highly tolerant to soil salinity, sodicity, acidity, Al, Mn and heavy metal (e.g. As, Cd, Cr, Ni, Pb, Zn and Cu). Both green house- and field-trials in Queensland, Australia have shown that vetiver is suitable for the rehabilitation of metal contaminated soils, and for the treatment of landfill leachate (Truong and Baker, 1998). Other

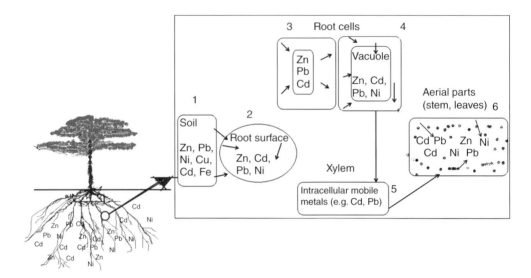

Figure 5.5 Metal uptake and accumulation in plants (Lasat, 2000). 1, Bioavailable heavy metals in soil; 2, Metal adsorption at root surface; 3, Movement of bioavailable metals into root cells; 4, Immobilization of metals into vacuole; 5, Movement of mobile metals into xylem and 6, Metal translocation into stem and leaves.

grass species (*Paspalum notatum*, *C. dactylon* and *Imperata cylindrica* var. *major*) have been used for revegetating Pb/Zn tailings in China and for coal mine wastes in Dhanbad, India. However, vetiver has superior biomass production and coverage (Singh et al., 2000; Shu et al., 2002; Tripathi and Singh, 2011).

In Australia, vetiver has been successfully used for stabilising overburden and highly saline, sodic, magnesic and alkaline (pH 9.5) and highly acidic (pH 2.7) arsenic-rich tailings, the latter from the mining of gold. In South Africa, vetiver has been used to stabilize 'slime dams', when ambient temperature exceeded 55°C (Truong, 2004).

In other applications, vetiver can be used to intentionally remove specific heavy metals from the contaminated sites, as their roots and shoots have been shown to accumulate more than five times the chromium and zinc levels in soil (Truong, 2004). A separate investigation at Tara mines in the Raniganj coalfields, India, indicated that the luxuriant growth of lemongrass (*Cymbopogon citratus*) and citronella grass (*Citronella winterii*) (Figure 5.6), was responsible for a reduction in total heavy metals found in the soil over time (Tripathi and Singh, 2011).

Vetiver's unique bio-engineering attributes arise from:

■ Its fast growing high-density root system that can grow >3m in length in the first year after planting,
■ Drought tolerance in dry soils and roots quickly reach the water table,

(a)

(b)

Figure 5.6 (a and b) Contaminated coal mine soil along with vegetative reclamation with aromatic grass.

- High tensile strength roots (ca. 75MPa) which can stabilise steep slopes by increasing soil shear strength by 40% at a depth of 0.5m,
- Close planting reduces water runoff mitigating soil erosion,
- Effective water filtering of surface run-off, and rapid root develop (from buried nodes) that keeps pace with ground level changes,
- Its tolerance to drought and extreme temperatures (14°C to 55°C) (Truong et al., 1996),
- Rapid regrowth after drought, frost, salt and other adverse impacts; and
- A high level of tolerance to soil acidity, salinity, sodicity and acid sulfate conditions (Le van Du and Truong, 2003).

5.3.5   Carbon sequestration

The natural transformation of atmospheric CO_2 into biomass through photosynthesis, and incorporation of biomass into humus accounts for the high levels of embodied carbon in soil. Soil contains approximately 75% of the terrestrial C pool – about 3.3 times more C than the atmosphere and 4.5 times more C than all living things (Houghton et al., 1985; Schlesinger, 1986; Lal, 2004; Washington Department of Ecology Report, 2015).

The combustion of fossil fuels and land use change including deforestation has lead to an increase in atmospheric CO_2 concentration from its pre-industrial level of approximately 280 to above 400 ppm (http://blogs.scientific american.com/observations/2013/05/09/400-ppm-carbon-dioxide-in-the-atmosphere-reaches-prehistoric-levels/). If this increase continues the atmospheric concentration will reach up to 720–1000 ppm by the turn of this century (Kumar et al., 2006; IPCC, 2014; Bierwirth, 2015).

Terrestrial ecosystems play a major role in moderating the global C cycle, but the human-induced perturbations do directly affect ecosystem functionality. Mining is an anthropogenic activity causing drastic soil disturbance, and when managed appropriately these soils have significant potential for locking up carbon, creating biological resources that sequester C and restore land to productivity.

Thus, a potential approach to the mitigation of CO_2 is via enhanced C sequestration in terrestrial ecosystems via an increase in biomass production and the creation of long-lived plant and SOM pools that are resistant to microbial decomposition. Underpinned by soil-plant-microbe relationships, plant- and soil-based C sequestration strategies can be designed to reduce net CO_2 emission into the atmosphere.

Owing to its large land area and diverse ecoregions, India has significant potential to capture carbon in soils, particularly via revegetated mine spoil. As the use of natural organic soil materials/amendments including peat, mulches, soil stabilizers, improves soil health and biomass productivity (Norland, 2000). By enhancing SOC pools in degraded soils, both soil quality and ecosystem productivity can be improved over long timescales (Lal et al., 1995, 1998; Akala and Lal, 2001; Tripathi et al., 2014). However, the development of SOC depends on the rate of formation of a suitable soil surface horizon that is rich in SOC, and this is influenced by cultural practices, land use and type of plant species.

In the United States, about 3.2 Mha are mine-related soils with a C sequestration potential rate of 0.5–1 T C ha^{-1} year^{-1} (Lal, 2000; Office of Surface Mining (OSM), 2003; Lal, 2004b). This equates to the sequestering of 1.6–3.2 MT C year^{-1} and is equivalent to the offsetting of 5.8–11.7 MT CO$_2$ year^{-1} from the burning of coal.

The estimated carbon sequestration potential of the world's agricultural soils lies between 0.4–1.2 Gt C/ year, while for other land uses (forests and agroforestry) the value could be 2.5 Gt C/yr (Lal, 2004b; Soussana, 2015). Agricultural soils have a technical carbon sequestration potential between 0.7 and 1.2 GtC/yr, while the potential from all other land uses (including forests and integrated systems like agroforestry) could reach 2.5 GtC/yr. Since 50% of a forest's standing biomass is carbon itself, trees are the very important sink for atmospheric carbon. It is further estimated that 2–4% of net fixed C in plants may be directly deposited into the soil via root exudates (Jones et al., 2004; Day et al., 2010), and this can be accessed by plants (and be uptaken) in a controlled way (Farrar et al., 2003).

Trees direct a greater proportion of their fixed carbon belowground than annual plants, with 40–73% of assimilated carbon being fixed (Grayston et al., 1997), (Fogel and Hunt, 1983). A varying proportion of C from photosynthesis is allocated to leaves, roots, storage, metabolism and root exudates but this greatly regulates soil C storage and varies on the environment the storage of C in soil (Epron et al., 2012). However, the amount of C 'stored' belowground ultimately controls the amount of C in a soil. As SOC enhances microbial activity, the presence of microorganisms can initiate a feedback loop that in turn increases root exudation (Meharg and Killham, 1991). Plant roots and grass species support intensed microbial metabolic activity, which releases C as CO$_2$ (Warembourg et al., 2003). Thus, respiration plays an important role in C storage in soil. The allocation of C in belowground biomass is shown in Figure 5.7. Carbon from plant roots, therefore, strongly influences the soil microbial community and, consequently, soil health (Brant and Myrold, 2006; Singh et al., 2012 Report). The carbon sequestration potential of the world's soils is given in Table 5.2.

5.3.5.1   *The mechanism of carbon protection*

The protection of soil C is an integral part of its sequestration in both the tropical and temperate ecosystems. Several mechanisms are identified for the protection of SOC (Greenland et al., 1992; Cambardella and Elliot,

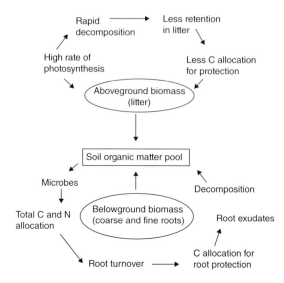

Figure 5.7 Allocation of C to belowground biomass (Grayston et al., 1997).

Table 5.2 Carbon sequestration potential of the world's soils.

Land use	Area (ha)	C sequestration potential (GtC year^{-1})
Cropland soils	1.35×10^9	0.40–0.80
Rangeland, grassland	3.70×10^9	0.01–0.30
Irrigated soils	275×10^6	0.01–0.03
Degraded soils/forest soils	1.10×10^9	0.20–0.40

Source: Lal (2004b).

1993; Palumbo et al., 2004; Bossuyt et al., 2005) that are chemically, physically and biologically influenced (see Figure 5.8).

Chemical 'protection' is achieved through bonding between minerals and formation of recalcitrant compounds, whereas physical 'protection' is rendered through the formation of soil aggregates. These soil micro-aggregates are particularly important to long-term C sequestration as they are resistant against decomposition, resulting in much longer residence times.

Soil aggregation results from the rearrangement, flocculation and cementation of soil particles, which is mediated by SOC and, biota, ionic bridging reactions, and the presence of clay and carbonate minerals (Bronick and Lal, 2005). Aggregation can be enhanced by root growth, fungal hyphae and mycorrhizae, and the development of SOC and these aggregates can encapsulate primary soil particles and release organic compounds, which, in turn, bind the particles together.

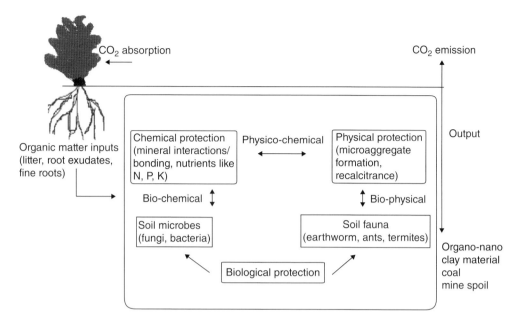

Figure 5.8 Factors affecting organic carbon protection and release (Chevallier et al., 2004).

The protection provided by micro-aggregates is due to the sorption of organic substrates onto clay mineral grains, particularly in 0.2–5 mm aggregates (Chevallier et al., 2004).

Thus, the terrestrial sequestration of C is reliant upon a plant-, microbial- and soil-based interactions to be effective.

Although mine spoil is are often characterized by an acidic pH, low concentrations of key nutrients, a poor soil structure, and limited moisture-retention capacities, as mentioned earlier, it also possesses a significant C sequestration potential (Akala and Lal, 2001).

5.3.6 Aesthetic enhancement

Re-vegetation enhances the aesthetic beauty of degraded land, whilst also improving drainage and slope stability. Vegetative cover also reduces potential dust hazards by disrupting wind at ground level.

The appropriate selection of plant species is the key factor for a successful reclamation action. In general, opportunistic plant species that germinate readily and grow rapidly to provide ground cover are ideal for re-vegetation. These plants are generally hardy and tolerate of extreme heat, frost and wind and promote the formation of humus. However, the survival of an individual species is not as important as the progressive establishment of naturally viable vegetative regimes. This can be achieved through the stabilization of an area with grass species (and other forms of rapidly colonizing ground cover), prior to establishing large shrubs and trees, with preference being given to endemic species for the area for the conservation of biodiversity.

5.4   Rebuilding soil structure

The structure of a reclaimed soil is very important to re-establishing a terrestrial ecosystem. The way soil particles are held together (i.e. by soil aggregation) and their size are particularly important. The establishment of native species formation of an 'A' horizon has been reported with organic matter 10 years after re-vegetation of mine spoil in Kentucky, however elsewhere, no visible changes in native species establishment have been observed even 30 years after re-vegetation (Johnson et al., 1982; Russel and La Roi, 1986).

The degree of soil compaction depends upon the methods adopted during soil replacement (Visser et al., 1984), and transporting. Transporting soil to a site by tractor or conveyor can help improve soil macrostructure by breaking up large aggregates/clumps, and by forming a 'skin' of fine particles to smaller aggregates, which promote a loose soil structure. Loosely constructed, or 'fritted', subsoil is very important for the proliferation of root systems, which, in turn, determines a plant's ability to maximize its surface area and access a greater volume of water and soil nutrients. Roots grown in fritted subsoils show extensive vertical and later penetration, which in turn improves soil properties in the formation of SOM and enhanced soil biological activity (Young, 1997).

Tree roots improve soil structure and functionality in several ways. One of the most significant plant-induced changes in soil structure by penetrating roots is the formation of channels via continuous macropores (Angers and Caron, 1998). These are formed by compressive and shear stresses when roots grow in a soil matrix (Goss, 1991). A large proportion of the pores formed fall into the macropore range (>30 µm) (Gibbs and Reid, 1988), facilitating soil aeration, water percolation and storage. These pores also create zones of failure, which help to fragment the soil, forming aggregates, further decreasing resistance to further root growth.

Radial pressure exerted by growing roots compresses soil (Dexter, 1988), enlarging existing pores and creating new ones. Roots can form channels through compacted soils vastly increasing water infiltration and transport efficiency (Barley, 1954; Yunusa et al., 2002; Bartens et al., 2008), although water flow may be greater after the death and decay of roots (Mitchell et al., 1995). In addition, fine roots play an important role in maintaining SOM and mineral nutrients through exudation and upon mortality (McClaugherty et al., 1982). Soil inhabited by plants tends to dry more quickly due to transpiration resulting in greater shear and tensile strengths and this leads to an increased resistance to root slip root/soil tangential resistance to slipping (Waldron and Dakessian, 1982). Lower soil water content may also help soils resist compaction (Horn and Dexter, 1989; Lafond et al., 1992). As the surface of a reclaimed mine spoil weathers its slope will decrease over time. As rock fragments weather and soil-sized particles are formed the water retention characteristics of spoil are improved and the growth potential of the plants is enhanced.

5.4.1   Recharging soil microbe activity

Microbial activity is a key regulatory factor of terrestrial ecosystems. The decomposition of biomass and the cycling of nutrients serve as the 'eye of the needle in re-establishing the soil functionality. Soil microbial biomass is a good indicator of the health of soil, and the metabolic activity of soil microbe populations can be used to determine the stability of an ecosystem, as it shows adaptation to change in ambient conditions (www.elib.edu.et/bitstream/123456789/41883/2/27324.pdf). In disrupted soil, microbial activity may significantly decline and be slow to resume, as bacterial species and fungi require plant material to support their symbiotic relationship.

5.4.1.1   *Bacteria*

When soil is excavated and stockpiled, the bacteria inhabiting the original upper soil layers are buried and compacted and this has an adverse impact on microbial activity. A flush of activity occurs in new upper soil layers during the first year as exposure to atmospheric oxygen increases (Williamson and Johnson, 1991), but after two years, less than one half the initial population persists at depths below 50 cm. As the reduction in oxygen availability can inhibit microbial metabolic processes, also a reduction in readily oxidizable carbon to fuel to stimulate nitrogen cycling, is also important. In this respect, the use of leguminous plants containing nitrogen-fixing bacteria may help 'recharge' a soil.

5.4.1.2   *Mycorrhiza*

Mycorrhiza is a mutualistic association between plants and fungi that affects the functioning of all terrestrial communities, improving the growth and fitness of introduced plant species.

In general, the majority of plants growing under natural conditions have mycorrhizae (Smith and Reed, 1997), and the colonization of roots increases root surface area increasing nutrient uptake. The association stimulates the uptake of phosphorus and nitrogen and as fungal hyphae can extend several centimetre (from the root) into soil the uptake of large amounts of nutrients is enhanced.

Mycorrhizal fungi have evolved a heavy metal tolerance mechanism and play an important role in the phytoremediation of degraded soil. Since heavy metal uptake and tolerance depend on both plant and soil microbes, information on the interactions between plant root and their symbionts, such as arbuscular mycorrhizal (AM) fungi, and nitrogen-fixing microbes is also important.

The hyphal mycorrhizal fungi network breaks when soils are initially moved and stockpiled (Gould et al., 1996). The viable inoculum potential slowly declines during the first 2 years of storage (Miller et al., 1985). The viability of mycorrhizae in stored soil does however decrease considerably thereafter to approximately 1/10 of undisturbed soil (Rives et al., 1980); partially as the amount of water in soil significantly affects mycorrhizal viability. When soil water potential is less than –2 MPa (drier soil), mycorrhizal propagules can survive for longer, and when greater than –2 MPa, the length of 'storage' time is important (Miller et al., 1985). In drier climates, thickly stockpiled mine spoil may not threaten mycorrhizal propagule survival. In wetter climates, shallow stockpiles are more important for survival, as the stockpile surface-to-volume ratios enhance moisture evaporation.

Mycorrhizal propagule density remains low immediately after reclamation on non-inoculated sites but re-establish themselves after a period of about 2 years (Gould et al., 1996). Lindemann et al. (1984) found that covering spoil with 30 cm of topsoil (without mycorrhizal inocula) stimulated host colonization, whereas hay, topsoil with inocula or sewage sludge had no effect. It is possible that sewage sludge may even suppress mycorrhizal development by increasing available phosphorus to host plants (Daft and Hacskaylo, 1976). Interestingly, the soil microbial populations persist in stored soil and can be stimulated during reclamation with a source of organic carbon or by adding suitable host plants.

Managing the microbial population in the rhizosphere by an inoculum consisting of a consortium of plant-growth promoting rhizobacteria, mycorrhiza-helping bacteria, and N-fixing rhizobacteria. The AM fungi can support the plant establishment and it is recommended to use indigenous strains in an inoculum, as these will be best adapted to the prevailing soil and climatic conditions.

5.4.2   Re-establishment of nutrient cycle

Mining activities alter the elemental distribution and, concentrations (and type), of metals in soils and this has an adverse on soil organisms.

Nutrient cycling involves an interaction between the inorganic elements (for example, CO_2, ammonia) and organic molecules inside organisms or detritus involves. In a managed ecosystem, nutrient cycling is very important to establish otherwise the supply of essential elements will restrict crop growth and establishment.

Nutrient cycling by soil microbial biomass involves the delivery and reuse of carbon, nitrogen, and phosphorus via the metabolic activity of plants and soil microbes. Carbon and nitrogen cycling is disrupted if soil microbial populations decline and a strong relationship is found between the time that has elapsed after restoration and an increase in soil microbial biomass.

5.4.2.1 Carbon cycle

The metabolic activity of soil microbes is fueled and regulated by availability of organic carbon (Williamson and Johnson, 1991). Microbes obtain carbon through their symbiotic relationships with host plants, and the availability of organic carbon in the soil, resulting from decomposition of plant and animal residues. The mixing of topsoil with subsoil at a mining site considerably reduces the relative available proportion of organic carbon (Visser et al., 1984), however amendments with bark (Elkins et al., 1984) or fertilizing with farmyard manure (Williamson and Johnson, 1991) can provide enough organic carbon to microbes to stimulate metabolic activity, which can be measured by increased microbial biomass carbon. Plant such as *D. sissoo* have been reported to improve the field moisture content (7%), pH (5.5), organic carbon (85%) and NPK through the accumulation of leaf litter and its decomposition to form humus (Maiti and Ghose, 2005). Revegetated mine spoil is recalcitrant at the early stage of planting, but the gradual accumulation of organic matter increases embodied carbon.

5.4.2.2 Nitrogen cycle

SOM has a very important influence on soil the physico-chemical properties and biological activities in soil. Nitrogen derived from organic matter is converted by micro-organisms into ammonium (NH_4^+) and under certain conditions, specific microbes use ammonical N for energy by oxidising ammonium $N(NH_4^+)$ first into nitrite $N(NO_2^-)$, and then into nitrate $N(NO_3^-)$, which can then be used by plants through nitrification. Free atmospheric nitrogen can in turn, be "fixed" by leguminous plants that will eventually be eaten or die, starting the cycle again. Davies et al. (1995) reported the nitrification rates in reclaimed sites were less than those in undisturbed sites during the first two years, but were similar after this time.

Nutrient recycling and availability on revegetated sites are reflected in part by the rate of decomposition of plant material, which is often retarded immediately after reclamation (Lawrey, 1977). However, as litter. Decomposition rates begin to equalize after 6 months

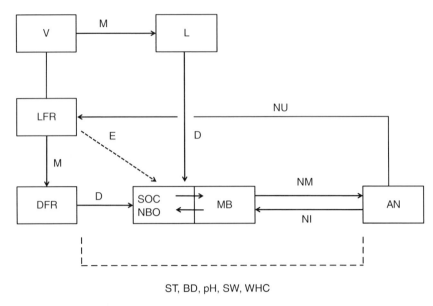

Figure 5.9 Box and arrow diagram representing major pools (boxes) and processes (arrows) involved in nitrogen cycling. AN, available (inorganic) nitrogen; D, early decomposition phase incorporating organic C and organically bound nutrients into soil; DFR, dead fine roots; E, exudation; L, litter; LFR, live fine roots; M, mortality affected by subsidence here; MB, microbial biomass; NBO, nitrogen bound to organic matter; NI, nitrogen immobilization; NM, nitrogen mineralization (including nitrification); NU, nitrogen uptake; SOC, soil organic carbon; and V, vegetation. Major soil physico-chemical variables that affect the processes are BD, bulk density; pH; ST, soil texture; SW, soil water; and WHC, water-holding capacity (Tripathi et al., 2012).

suggesting the establishment of 'normal' microbial activity follows. Amendments involving bark are known to significantly increase soil microbial activity (and consequently decomposition rates), but result in less-available NO_3^- (Elkins et al., 1984). Plant litter and root exudates provide establishment of nutrient cycling. Figure 5.9 shows the cycling of N in mine soil.

5.5   Determining the effectiveness of soil reclamation

The reclamation of degraded land is a very complex process. Once new vegetation is established the effectiveness of the approach adopted can be evaluated by determining the state and functionality of the soil system. As there is no single indicator of the health of an ecosystem, the activation of soil biological processes and the formation of

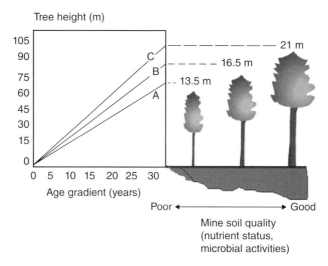

Figure 5.10 Tree growth as a function of mine soil quality and quantity. Relative stem value increases exponentially with tree height as mine soil quality/quantity increases (Burger, 1999).

stable aggregates are key to soil functionality (Filip, 2002; Sourkova et al., 2005; Heras, 2009).

In addition to establishing a commercially valuable resource of trees that sequester atmospheric carbon, the Indian Forestry Reclamation Approach allows the return of the full forest ecosystem succession. Established forests stabilize mine soils, maintain clean drinking water, provide watershed protection services, enhance wildlife habitat and improve landscape aesthetics. Landowners benefit from a post-mining timber resource of significant economic value, whereas mining companies benefit from reduced reclamation costs. The return of wildlife, watershed protection and increased carbon storage in the soil are additional benefits.

As illustrated in Figure 5.10, Burger (1999) demonstrated that trees grow slowly on poor-quality mine sites and by age 25 are shorter (site index = 13.5 m) then trees growing on average mine soil quality (site index = 16.5 m) or trees on good mine soil with improved nutrient status (site index = 21 m).

5.6 Costs of bio-reclamation and employment generation

The reclamation of mine waste creates employment opportunities for local people and it is estimated that more than 400 man-days/ha can be available to local people on the short-term basis (Table 5.3). The cost of reclaiming 1 ha, including five years of maintenance

Table 5.3 Average man-days generated during reclamation of 10 ha area.

Item	Man-days	Cumulative man-days	Remarks
First year			
Pit digging	1000	1000	25 pit/man/day and a total of 2500 pits/ha
Plantation of sampling	500	1500	Planting of 50 sapling/man/day and a total of 2500 plants/ha
Irrigation	260	1760	An average 1 irrigation per week and 1 person/ha for approximately 6 months
Maintenance	520	2280	An average 1 maintenance per week and 1 person/ha
Guarding	365	2645	One guard for the whole area
Second year			
Irrigation	130	2775	An average 1 irrigation per week and 1 person/ha for 3 months (dry season)
Maintenance	260	3035	An average 1 maintenance per week and 1 person/ha for 6 months
Guarding	365	3400	One guard for the whole area
Third year			
Irrigation	130	3530	An average 1 irrigation per week and 1 person/ha for 3 months (dry season)
Maintenance	130	3660	An average 1 maintenance per week and 1 person/ha for 3 months
Guarding	365	4025	One guard for the whole area

Table 5.4 Suggested methods of re-vegetation with cost.

Types of vegetation	Pit size (m)	Topsoil amendment	FYM (kg/pit)	Bio-fertilizer	Cost/ha (Rs)
Fruit trees	0.5×0.5×0.5	25% of pit size	≥10	*Rhizobium* *Azotobacter* Mycorrhiza	40000
Trees, shrubs and grasses	0.5×0.5×0.5 0.3×0.3×0.3 0.1×0.1×0.1	25% of pit size	5–6	*Rhizobium* *Azotobacter* Mycorrhiza	30000
Green belt	0.5×0.5×0.5	25% of pit size	5–6	*Rhizobium* *Azotobacter* Mycorrhiza	30000

(for various wasteland, slime, saline land, alkaline land, ravine land and hydro-seeding) are presented in Table 5.4.

6 Best practice reclamation of mine spoil

6.1 Background

Feasible solutions for mitigating the impacts of mining involve using a best practical and economic solutions-approach and can be simply explained as "the best way of doing things". Environmental impact mitigation demands a continuing, integrated process extending through all project phases, from mineral exploration to mine closure and will:

- Decrease pollution loading, via reduced availability and concentration
- Reduce erosion and sedimentation problems, and
- Improve the productivity of abandoned mine land

The successful reclamation of mine land may involve combining technologies in a 'treatment train', to re-instate forest, pasture, or a habitat that supports the former, primary land use. This outcome, however, depends partly on the location of a site and the condition of the mine soil, and the nature of surface and sub-surface waters. In general, the success of reclamation depends on managing a range of factors and does not allow for a quick fix'.

Restored land requires post-closure management to establishment a sustainable ecosystem, and these activities require resourcing to the outcome is successful.

6.2 Soil management practices

The successful restoration of degraded mine soils depends on a site-specific approach and must consider site topography and future change

Reclamation of Mine-Impacted Land for Ecosystem Recovery, First Edition. Nimisha Tripathi,
Raj Shekhar Singh and Colin D. Hills.
© 2016 John Wiley & Sons, Ltd. Published 2016 by John Wiley & Sons, Ltd.

resulting from ground settlement and the steepening of slopes and their effect on slope instability.

Overburden may settle over time, causing depressions that can become waterlogged. Change in surface materials (both top and sub-soil), increases the potential for erosion, as soil permeability, structure and infiltration capacity are impaired.

6.2.1   Topography and soil erosion

Generally, surface mine operators revegetate with a grass or shrub species to minimize run-off on gentle slopes (usually below <40%). These 'gentle' slopes have reduced erosion potential, require minimal slope-regrading and can be deep tilled or ripped. On these slopes, the infiltration rate will be similar to level ground, and general farm tillage can still take place.

If barren mine spoil are level, there is reduced run-off and also fewer requirements for water control. Differential settlement across the disturbed land is addressed via periodic levelling, and the installation of subsurface drainage. Soil erosion can be checked via cover crops, and legume, and grass cover may help build organic matter and to reduce soil erosion in the short term.

On longer gentle slopes, drainage control structures are recommended. The gradual or steeper slope may require terraces or water and sediment control basin (WASCOB) structures for soil stabilization.

Each method must be checked and maintained after monsoon rain and large storm events.

6.2.2   Compaction and bulk density

The movement of heavy vehicles, machinery and mine equipment increases the compaction/bulk density. The degree of compaction is, however, affected by soil texture and its moisture content during dumping operations, and can significantly impact on water infiltration and plant root penetration and proliferation.

To manage compaction of spoil, the movement of heavy vehicle and machinery should be minimized, and planting is recommended before the onset of winter weather, or monsoon conditions, thereby avoiding further disturbance.

6.2.3   Deep ripping/tillage

During mining operations, soil compaction can be high due to large equipment and heavy vehicle movements. Ripping and tilling can be used to improve water infiltration rates by increasing the amount of macropore space in soil. It is not known, however, if these benefits are long lasting.

Deep chiseling (similar to ripping but shallower) is another effective mechanism to prevent run-off and erosion on relatively flat slopes (Verma and Thames, 1978). Chiseling can be used in conjunction with other treatments in mine spoils. In general, however, shallower ripping (18–24 cm) can be used to alleviate compaction of surface layers, whereas deep tillage/ripping, 60–100 cm, can also be beneficial.

6.2.4 Ground cover residue management

To combat the effects of wind and water erosion, residue management at ground level can be helpful. Conservation tillage involves using the previous year's crop residues (such as stalks or stubble) as cover when planting the next crop, as this helps reduce runoff. A minimum or conservation tillage approach protects ground and surface mulch, applied to 30-70% of the area being re-vegetated protects soil from wind and water erosion, and enhances moisture and nutrient retention capacity. Additional benefits from this approach include minimal soil compaction or crusting of the soil surface.

Residue retention with 'no-tilling' or 'less-tilling' enhances the storage of organic carbon in soil. For the early establishment of a plant species, one of the following tillage practices should be used:

- *No-tilling and strip-tilling* distributes crop and other plant residues on the soil surface throughout the year. Planting is carried out directly in narrow tilled slots or directly in untilled soil. Seedlings/ saplings are placed in a narrow seedbed, where no more than one third of the row width is disturbed.
- *Mulch-tilling* involves that the entire field surface and regulates the amount, orientation and distribution of crop/plant residues on the soil surface prior to planting. This is the reduced tillage system that leaves at least one third of the soil surface covered with plant/crop residue.
- *Ridge-tilling* pre-forms ridges on which plants are grown (100– 150 mm high) alternate with furrows protected by plant residue. The soil is left undisturbed after harvesting as the previous crop residues are cleared off ridge-tops into the adjacent furrows (to reduce moisture loss) and make way for the new crop. The maintenance of the ridges is essential and requires modified or specialised equipment.

6.2.5 Water management

To enhance water infiltration and movement of soil during a monsoon, conservation practices can be implemented in a single or combined approach. The root zone available water capacity i.e., the capacity of a soil to hold water in a plant-available form (expressed in cm of water per cm of soil, multiplied by soil rooting depth) determines the productivity of reclaimed soils. This is a key property and is important in developing water budgets, predicting potential periods of drought, designing drainage systems, and the likely plant biomass productivity (Tripathi and Singh, 2008). Soils with low root zone available water are likely to be dry and less supportive of plant growth. Soil water conservation management practices for maintaining/increasing plant growth include:

Residue/organic matter management: It is important to manage the amount, orientation and distribution of plant and other plant residues on the soil surface year-round. This practice may be applied as part of a conservation system to:

- Reduce soil erosion
- Maintain and improve soil organic matter content
- Conserve soil moisture
- Manage dew to increase plant-available moisture
- Reduce off-site transport of sediment, nutrients or pesticides

Covering early revegetated land with polyethylene sheeting with holes for saplings checks evaporation of water from soil and accumulates dew at the bottom of the plants.

Terracing: Soils subject to erosion and excess water runoff require attention to conserve water by terracing. Terracing is a soil conservation practice for sloping land and uses earth embankments or a combination of terraced ridges and channels constructed across the field slope. In mine spoils, terracing can be efficiently used to:

- Conserve water
- Check excess run-off of water; and
- Provide a suitable water outlet

Contouring: The use of contour tillage involving planting and other farming operations on or near the contour of the field slope to:

- Reduce soil erosion
- Reduce transport of sediment and other waterborne contaminants
- Increase soil infiltration

This practice is used on sloping land where there is no contamination and is most effective on slopes between 2 and 10%, but not suitable for slopes exceeding 10%. Various contouring techniques used in mine reclamation are seen in Figure 6.1.

WASCOB: These are earth embankments or combinations of ridge and channel a constructed across a slope on minor watercourses. These are effective at trapping sediments, retain rainwater and improve water quality. WASCOB are used to control erosion in small drainage channels, whereas terracing controls water flow on slope surfaces. Suitable sites for WASCOB are where:

- The topography is generally irregular
- Watercourse or gully erosion is a problem
- Sheet and rill erosions are controlled by other conservation practices
- Run-off and sediment damage land
- Soil conditions are suitable; and
- Adequate outlets can be provided

Conservation tillage: Conservation tillage is defined as a tillage and planting technique for establishing crops in the previous crop's residues left on a soil surface. Generally, conservation tillage, which leaves at least 30% of the field surface covered with crop residues after planting has been completed. The benefits of conservation tillage are to:

- Increase plant residue and soil organic matter
- Enhance plant-available moisture
- Improve soil quality

(a)

Figure 6.1 Contouring techniques used in mining areas. (a) A retaining wall along haulage road at Sirmour, HP, India.

(b)

Figure 6.1 (*Continued*) (b) Installation of catch dams at high-altitude limestone mines of Sirmour, HP, India.

- Increase earthworm populations
- Improve soil structure
- Increase infiltration rates, maintain carbon in soil and reduce water and wind erosion

Conservation tillage takes longer to impart change to the condition of soil and in crop response.

Grassed waterway: Grassed waterways are the natural or constructed broad and shallow channels designed to transport surface water at safe velocities. A grassed waterway consists of a green belt of native grassland strip (2–48 m), and the vegetative cover slows

(c)

Figure 6.1 (*Continued*) (c) Installation of catch dams and vegetative reclamation at high-altitude limestone mines of Sirmour, HP, India.

water flow to protect the surface. This practice may be applied as part of a resource management system to support one or more of the following purposes to:

■ Check run-off without causing erosion or flooding
■ Reduce rill and gully erosion
■ Protect/improve water quality

Filter strip: Filter strips are narrow and long vegetated areas (1–1.5 m wide) between surface water bodies (such as wetlands, streams and lakes) and cropland, grazing land, forestland or disturbed land. They are also known as vegetative filter strips or buffer strips and slow the rate of run-off, reducing sedimentation and the run-off pollutants (e.g. phosphorus, nitrogen and pesticides).

   This approach can also lessen sheet and rill erosion, improve soil aeration, lessen water quality degradation (by nutrient removal in the root zone by plant uptake and sorption to soil), and provide wildlife habitat. Filter strips are most effective at slopes between 2 and 6% because of the increased contact time between the run-off and the filter strips. The ecological benefits of using filter strips include:

- Soil is stabilized by roots via soil aggregation
- Vegetation absorbs the force of wind, water and the impact of rain drops
- The soil moisture content is improved
- The strip acts as a noise filter and attenuator; and
- Nutrients are recycled, reducing crop and animal stress resulting from hot/dry summers and cold winter winds.

Limitations do include the cost of installation (e.g. grading slopes and vegetation establishment), weed control, the cost of maintenance and the effectiveness to extreme rates of run-off rate and frequent events.

6.2.6   Woodland management

Revegetated plant and tree species established on mine spoil can be managed in the same way as those on virgin land (Richards et al., 1993). However, individual site circumstances may require additional intervention such as the:

- Application of nutrients after planting to retain nitrogen-fixing nurse species (such as Crotolaria juncea)
- Selective management of nurse-species to ensure species variety
- Appropriate control of grass and scrub to reduce the risk of fire
- Grazing of young woodlands after 4–5 years; and
- Application of sewage sludge (as a source of nutrients and organic matter) to facilitate woodland development

To minimise the costs of managing woodland, native timber species should be planted as crops for manufactured products, and the production of woodchip, mulche and compost.

6.2.7   Practices to enhance carbon sequestration

Soil carbon content is defined as persistent increase in SOC originating from atmospheric CO_2 (Torri et al., 2014). Sequestrated carbon increases with the SOC pool, as re-vegetation stimulates succession and enhances biomass and soil productivity. Additions of organic amendments and 'green' management practices also help accelerate this process. Thus, by improving soil SOC in degraded system, the retention of carbon can be enhanced and an additional 1 Mt of carbon can be removed from the atmosphere each year (Pacala and Socolow, 2004, Kumar et al., 2015).

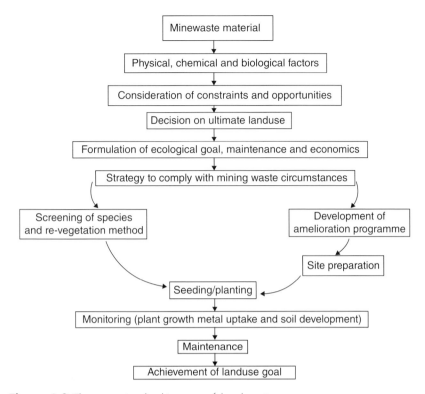

Figure 6.2 The stages involved in successful reclamation.

It is estimated that approximately 600 Mt to 1 Gt of carbon can be sequestered by the restoration of degraded soils every year (Lal and Bruce, 1999; Dulal et al., 2011). As the presence/addition of organic matter encourages greater carbon accumulation (Tian et al., 2009), reclamation practices employing organic amendments have benefits that include longer-term greenhouse gas avoidance (Brown and Subler, 2007).

A flow chart for generalized mine spoil reclamation steps is given in Figure 6.2.

7 Carbon uptake into mine spoil in dry tropical ecosystems

7.1 Background

Since the beginning of the industrial revolution, the concentration of CO_2 in the atmosphere has been steadily rising now exceeds 400 ppm (www.http://co2now.org/, accessed 21 March 2015). The increased use of fossil fuels and changes in land-use augmented by deforestation are key drivers of climate change (Keeling and Whorf, 2004).

Over the period 2005–2014, the average annual increase of CO_2 in the atmosphere was 2.11 ppm year^{-1} (Earth's CO_2 home page, 2014), and if this rise continues, atmospheric CO_2 will reach >1000 ppm by 2100 (Kumar et al., 2006; IPCC, 2015). The impact of climate change is wide-ranging, and the Doha Climate Change Conference (26 November–8 December 2012) recognized that losses in developing countries likely to be greater than for developing countries (Earth Negotiations Bulletin, 2012) may be relatively greater than for developed countries.

Coal is the single largest single energy source, providing about 40% of the electricity produced worldwide. India produces 63% of its power from coal. About 5.4 Gt of coal is burnt pa worldwide generating about a third of the world's CO_2 emissions. Globally, approximately 32,579 Tg of CO_2 were added to the atmosphere through the fossil fuel combustion in 2010, of which the United States accounted for about 17% (EIA, 2012).

Global CO_2 emissions from fossil fuel combustion reached a record high of 32.2 Gt in 2014, albeit unchanged from the preceding year (EIA, 2015). Although in 2012 CO_2 emissions declined in the OECD countries, the total energy-related CO_2 emissions increased by about 1%. The New Policies Scenario of World Energy Outlook (WEO, 2013) reports an unabated growth in global CO_2 emissions from fuel combustion, reaching 37.5 Gt CO_2 by 2035, leading to a long-term temperature increase of 3.6°C, vastly exceeding the 2°C target agreed at the UNFCCC (IEA, 2013).

Reclamation of Mine-Impacted Land for Ecosystem Recovery, First Edition. Nimisha Tripathi, Raj Shekhar Singh and Colin D. Hills.
© 2016 John Wiley & Sons, Ltd. Published 2016 by John Wiley & Sons, Ltd.

About 45% of the energy-related emissions are resulted from coal, followed by oil (35%) and natural gas (20%). The anthropogenic perturbations of the global C cycle not only directly affect global climate but also the function of ecosystems. In India, however, its large coal reserves and the low production costs encourage coal use; as such 18 states currently produce coal and lignite (Figure 7.1).

The world's soils contain approximately 1500 Pg ($1\,Pg = 1\,Gt = 10^{15}\,g$) of organic carbon (Batjes, 1996), roughly three times the amount of carbon in vegetation and twice the amount in the atmosphere (Intergovernmental Panel in Climate Change (IPCC), 2001; Denman et al., 2007; Scharlemann, 2014). The annual fluxes of CO_2 from the atmosphere to land (global net primary productivity, NPP) and land to atmosphere (respiration and fire) are of the order of $60\,Pg\,C\,year^{-1}$ (IPCC, 2000).

Mining causes drastic change to soil leading to the loss of SOC. Normally, during mining 30 cm topsoil is removed and stored separately (although this is not a common practice in India) and the overburden (covering the top) is then excavated and backfilled into an already mined area of the pit. During reclamation, the overburden is graded and topsoil is placed on top to a depth of 30 cm. The topsoil is then graded to the original contour of the land, and an initial dose of fertilizer/mulch is applied before seeding with a mixture of grasses and/or legumes.

In disturbed habitats, natural recovery takes much longer as colonization by plant and animal species is retarded. Man-made plantations exert a catalytic by changing sub-surface microclimatic conditions (viz. increased soil moisture, reduced temperature), increase vegetation–structural complexity and the development of litter and humus layers, particularly during the early years of plant growth (Singh et al., 2002).

Plant species, capable of surviving extreme conditions may be used as indicators of mine spoil/overburden chemistry/productivity, as they enhance the rate of soil mineralisation and surface carbon accretion. Re-vegetation of mine soils leads to a source of biomass, which in turn results in the increased availability of SOC to an ecosystem. During natural recovery, biomass accumulation takes longer (e.g., up to 12 years), compared to revegetated, amended soil (Jha and Singh, 1992). The accumulation of SOC could also be counted as an offset against CO_2 emission from coal mining (Shrestha and Lal, 2009).

Plantations stabilize soil through the development of extensive root systems. The re-vegetation of mine spoil can restore the rate of nitrogen transformation, increase soil microbial and plant biomass (Tripathi and Singh, 2008).

The age of revegetated mine soil is important, as for example, mature trees sequester more carbon than young trees. Soil organic

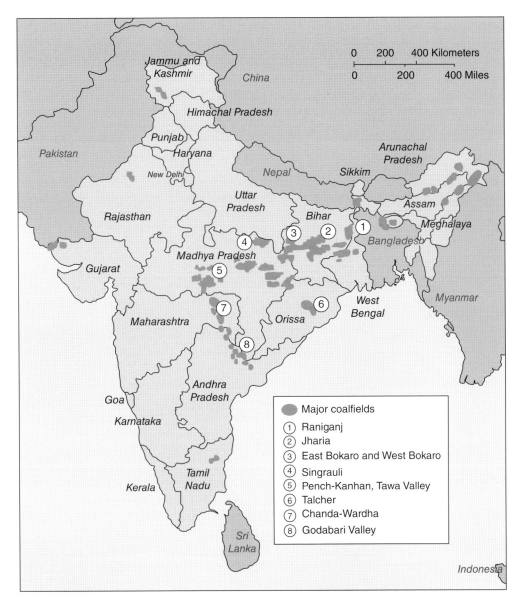

Figure 7.1 Coalfields within States of India (Trippi and Tewalt, 2011).

carbon from woody vegetation is more recalcitrant than that from herbaceous vegetation, due to higher concentrations of aliphatic suberin, waxes, glycerides, lignin and tannins in their roots (Liao et al., 2006; Lorenz et al., 2007; Filley et al., 2008; Dean et al., 2012). Dean et al. (2012) reported a 95% increase in the carbon in woody trees compared to herbaceous plants, as the stock of SOC is higher due to their deeper and more substantial roots.

According to Körner (2000), the majority of global vegetative carbon is stored in forests. If growth mass rate in trees is higher, the development of root mass and their hardiness are of prime importance to their ability to store terrestrial carbon.

The re-vegetation of mine spoil is the initial stage of soil development and offers a great opportunity to recycle organic waste material and maximize the amount of C which can be stored. Therefore, by appropriately managing the reclamation process, soil can be restored to a stable and productive state and in which large amounts of atmospheric C are stored (Shrestha and Lal, 2009).

7.2   Soil carbon sequestration

Soils are the largest terrestrial reservoirs of carbon, containing about three times more than is stored in vegetation, and twice as much as in the atmosphere (750 Pg of C) (USEPA, 1995; Batjes, 1996; Krishan et al., 2009).

The total SOC pool in Indian forest soils is estimated to range from 4.13 Pg C (for top 50 cm) to 6.18 Pg C (for the top 1 m) (Chhabra et al., 2003).

As soil degrades, it is estimated that each tonne of SOM releases 3.667 t of CO_2 to the atmosphere. Conversely, the formation of 1 t SOM removes 3.667 t of CO_2 from the atmosphere (Bowen and Rovira, 1999) and is retained through dynamic processes that balance additions and losses over time.

Terrestrial ecosystems are generally large C sinks supported by photosynthesis and the formation of complex carbon-based molecules. The bound energy (originally from sunlight) is transferred to minerals in soils in the form of litter, root turnover and exudates forming an intricate detrital trophic structure with in soil.

Carbon represents nearly half of the mass of live biomass and detrital OM in soil. The decomposition of OM completes the atmosphere–soil–atmosphere C cycle by emitting a portion of SOC back to the atmosphere as CO_2, while a portion is also incorporated with in the stable soil humus fraction. Decomposition rates tend to be proportional to the amount of OM present (Johnson, 1995), and over time and under relatively constant environmental conditions, an equilibrium is established between the rate of carbon addition and emissions resulting in the stabilization of SOC within a soil.

Mining and associated practices disrupt the SOC equilibrium, causing its loss. Good management practices help to regain carbon and improve soil quality. In natural habitats, carbon from plants is deposited in the soil through roots and plant residues, but its storage potential is determined by microclimatic, soil physico-chemical properties and biological activity. Once an equilibrium is established in treated mine spoil, it may contain less or more carbon than before mining took place, and this carbon may be stored belowground for centuries (Parton et al., 1987; Buyanovsky et al., 1994).

The mechanisms responsible for stabilizing SOC may be categorized as (Christensen, 1996):

■ Physical protection
■ Biochemical recalcitrance
■ Chemical stabilization

Chemical stabilization results from associations between decomposable organic and soil mineral components (e.g. organic carbon sorbed onto clay surfaces by polyvalent cation bridges or by trapping between expanding layers of clays).

Recalcitrant biochemical carbon is 'controlled' by chemical characteristics of C-containing substrate, such as lignin and its derivatives or fungal melanins (Haider and Martin, 1981). The nature of various organo-mineral associations and their distribution within soil aggregates determine the extent of physical protection and chemical stabilization of SOC (Gjisman and Sanze, 1998).

These mechanisms also play major roles in stabilization and retention of SOC in mine soils. A study across a chronosequence of 15 Montana surface mine sites investigated the recovery of carbon in the soil surface horizon (0–10 cm) into undisturbed levels within 30 years of reclamation (Schafer et al., 1980). Similarly, Illinois coal mining spoils spanning from 5 to 64 years in age have shown an increase in soil carbon by 0.1 and 0.3 Mg ha^{-1} year^{-1} in the upper 10 and 50 cm^2, respectively (Thomas and Jansen, 1985).

7.3 Carbon allocation in woody plants

All organic carbon found in the soil is primarily plant derived. The two main sources of carbon in the soil are:

(1) Accumulation of soil OM due to the humification after plant death
(2) Root exudates and other root-borne organic substances released into the rhizosphere during plant growth, as sloughing of root hairs and fine roots by root elongation.

The first mode of carbon uptake is well documented, but carbon sequestration by plant roots is still being investigated. Carbon dioxide is fixed by crop plants and is translocated into the roots in a simultaneous process. Carbon is transferred to soil by plant roots through their death, exudation and respiration; however, it is difficult to quantify the contribution of these three 'mechanisms' separately (Paul and Clark, 1996).

Wood is comprised of four basic materials, namely, cellulose (50% or more of the bulk), hemicellulose (20%), lignin (25%) and resin or oil (3–5%) (Moll and Moll, 1994). About 50% of the aboveground woody biomass weight accounts for carbon content in plants (Brown, 1997). Carbon storage varies in plant tissues with more carbon being in stems and fruits, than leaves. Moreover, longer-lived trees with high-density wood store more carbon per volume than short-lived, low-density, fast-growing varieties (Moura-Costa, 1996).

Woods contain suberin, waxes, glycerides, lignins and tannins. These compounds are recalcitrant and have a higher SOC sequestration potential. Moreover, the time required for 95% soil C sequestration by woody vegetation together with its deeper roots is longer, and the resultant SOC stock is higher (Liao et al., 2006; Lorenz et al., 2007; Filley et al., 2008). In woody biomass, the rate of carbon sequestration is regulated by two fundamental and dynamic processes, namely, plant growth and death (Dixon et al., 1994; Bonan, 2008; Purves and Pacala, 2008).

Carbon storage in a forests is allocated into five components: living trees, down dead woods, understorey vegetation, forest floor and soil and all are carbon storage pools (Birdsey and Heath, 1995). The carbon flux in living trees accounts for 76–90% of total carbon stored in a forest ecosystem. Studies show that the organic carbon stock in plant biomass was higher in a mature forest than in a recovering forest (Liu et al., 2006). Kaewkrom et al. (2011), however, suggested that the annual C accumulation rate in a mature forest may be less than in a recovering forest, as the former has a quite low growth rate compared to the later.

Climate, soil conditions and species selection have major influence on carbon stored in soil. To offset 1000 t CO_2e of greenhouse gases (over 30 years) in NE Australia, 1–45 ha land area was required for new plantation (Grace and Basso, 2012). The rate of biological carbon stored for a particular tree species depends on both site (soil fertility and water-holding capacity, etc.) and the prevailing climatic conditions. However, it is clear that a better understanding of what influences carbon storage in soil is required to optimise C storage in re-vegetated mine soils (Figure 7.2).

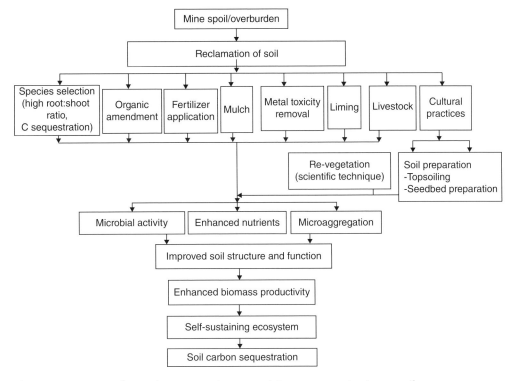

Figure 7.2 Factors affecting biomass productivity and C sequestration leading to self-sustaining ecosystem in drastically disturbed soil.

7.4 Mine spoil

During surface mining, the overlying soil is removed, and rock debris is deposited in the form of overburden. In this process, the land surface is completely stripped of vegetation, as overburden to access minerals or coal.

Surface mining imparts disturbance and degradation over large land areas and generates large volumes of heterogeneous material, comprising freshly blasted overburden or spoil, and once placed it can be reclaimed and utilized to support plant growth, either with or without topsoil (Sencindiver and Ammons, 2000).

Billions of tonnes of mine waste are produced annually (Bell, 1998). As we have seen, they are mostly too coarse, compacted and alkaline in nature and have low OM content (Tripathi and Singh, 2008), and do not readily support.

Plant growth, (Jha and Singh, 1992). Therefore, mine spoil is not easily colonized by plants or microbes as it is largely devoid of soil biota and

OM and in plant-available nutrients. Spoil often contains excess amounts of heavy metals (Singh et al., 1996).

The successful development of a soil may be the single most important edaphic factor determining the efficacy of a reclamation plan. Surface environmental conditions typically lead to rapid initial surface weathering, which decreases with time (Struthers, 1964). Rapid chemical reactions occur in spoil as the exposed material equilibrates with its new environment (Struthers, 1964). Soils are disrupted during weathering (by physical and chemical break down) and by the presence of roots. The availability of SOC to biomass and on root development is also influenced by the action of weathering (Haering et al., 1993). The decomposition of SOM and the aggregation of soil particles into peds are major pedogenic processes (Roberts et al., 1988a, 1988b), and are also accompanied by changes in the particle size distribution of soil particles.

Mine soils are pedogenically young soils, developing from fragmented and pulverized rock material and typically consisting of 32–67% rock fragments, which, is much higher than for native soils (Ashby et al., 1984; Ciolkosz et al., 1985; Thurman and Sencindiver, 1986).

Mine spoils offer very harsh conditions for both plant and microbial growth. Organic matter contents are low, and key physico-chemical characteristics are far from ideal (Singh et al., 2002), particularly, including: high pH, cation exchange capacity (CEC), bulk density and structure (Shrestha and Lal, 2009).

The coarse particle fraction of a typical mine spoil varies between <30 to >70% w/w, resulting from, blasting and handling techniques (Maiti and Saxena, 1998; Maiti and Ghose, 2005). In general, soils with stone content greater than 50% w/w are rated as poor (Hu et al., 1992), however, this will reduce overtime due to weathering.

The poor quality of mine spoils inhibits soil-forming processes and plant growth. The lack of SOM-associated nutrients and the paucity of 'topsoil', adversely influences fertility and productivity. Plant growth is limited by a low pH where of overburden contains sulphur-bearing minerals, such as pyrite, FeS_2 (Mays et al., 2000). The successful restoration of these soils presents a challenging task, as acidic.

Conditions limit plant growth, by interfering with nutrient uptake and root development. The use of lime to raise pH above 5.5 can be used to improve plant growth. During re-vegetation, overburden can be graded and mixed with farmyard manure before planting. Bradshaw (1983) recommends a minimum amount of total N (1000 kg ha^{-1}) to support plant growth, however, the availability of P may be limiting as it is often immobilized by freshly of exposed spoil (Bradshaw, 1983; Roberts et al., 1988b).

As mentioned, the use of heavy earth moving equipment results in the soil compaction and reduces soil air content/porosity (McSweetney and Jansen, 1984; Chong et al., 1986). Compaction leads to a higher bulk density and reduced hydraulic conductivity/water infiltration rates (Indorante and Jansen, 1981; Fehrenbacher et al., 1982; Thompson et al., 1987) and must be controlled (Thurman and Sencindiver, 1986; Shukla et al., 2004).

The bulk density of productive natural soil generally ranges from 1.1 to 1.5 Mg m^{-3}, while mine soils are often greater than 1.6 Mg m^{-3} (Daniels and Zipper, 1997; Daniels and Amos, 1981). Severely compacted mine soil (>1.7 Mg m^{-3}) with less than 0.6 m of effective rooting depth, cannot hold enough available water to sustain vigorous plant growth (Daniels and Zipper, 1997). A loose, non-compacted soil of 0.9–1.2 m depth will generally hold enough water to sustain plants during prolonged drought (Daniels and Zipper, 1997).

The loss of SOM results in poor soil quality as structure and function are impacted (Lal, 1997). Loss of SOM can be attributed to a lack of plant litter input and topsoil via the mechanical mixing of the A, B and C horizons (resulting from handling of overburden) Erosion, leaching and the accelerated decomposition of SOM in exposed topsoil are additional factors. The relatively rapid decline of the SOC pool in mine soil can be substantially reversed, as temporal changes in SOC enhanced soil quality and the establishment of plant species with improved structural and functional attributes.

7.5 Role of mine soil properties on C sequestration

Soil properties influence the rate of carbon uptake and its mineralisation. A soil's clay content and its water-holding capacity (and water content) influence the decomposition of organic matter, which is slowed down and may be incomplete, if soil is waterlogged (Rice, 2002); http://www.geotimes.org/jan02/featurecarbon.html).

pH is a measure of soil 'quality', which can change rapidly as rock fragments weather and oxidize, particularly if they contain pyrite. Carbonate (Ca/MgCO$_3$)-bearing rocks tend to increase the pH as they weather (Sheoran et al., 2010).

Soil pH impacts nutrient availability/fertility and influences the success of re-vegetation efforts. Acidic conditions mobilise metals and their toxicity potential can restrict plant growth. Fertilization and liming can raise soil pH above 5.5 to improve plant growth. A soil pH of 6.0–7.5 is ideal for forage and other agronomic or horticultural uses (Gitt and Dollhopf, 1991; Gould et al., 1996, Tripathi and Singh, 2008; Tripathi et al., 2012).

Soil texture is also important as aggregation influences hydrology and the availability of nutrients (Lindemann et al., 1984; Heras, 2009), but also reduces the potential for soil erosion (Elkins et al., 1984) and increases the stabilisation of soil carbon (Six et al., 2002).

Silt and clay may be bound to OM in soil to form organic-mineral complex. The close relationship between soil structure and soil organic carbon (Han et al., 2010) involves two processes: (i) organic carbon is chemically bonded to clay surfaces, slowing degradation. Clays with high adsorption capacities, such as montmorillonitic, can retain organic molecules and (ii) soils with higher clay contents have a potential to form aggregates, which trap organic carbon and physically protect it from microbial degradation (Rice, 2002, http://www.geotimes.org/jan02/featuree_carbon.html).

The particle size distribution of loamy soils is generally ideal for trapping SOM as is common where spoil comprises siltstones (Ghose, 2005). Aggregate structures can break down during soil excavation and stockpiling.

Almost all of the organic carbon in soil is located within the soil pore (between minerals) either as discrete particles or by chemisorptions onto mineral surfaces (Krull et al., 2001). Soil architecture influences the availability of oxygen and water and therefore, the stability of organic materials.

Plant–microbe interactions and the aforementioned C, N cycles also play a major role in soil carbon uptake (MacDonald et al., 2008). Plants provide carbon for microbial activity and its retention through litter fall and rhizodeposition. Microbial activity directly affects SOC and indirectly influences plant carbon accumulation via the N cycle. The supply of N involves complex processes governed by nature of the SOM (i.e., its C:N ratio), microbial nutrient mineralization activity and the total amount of N, which results from complex biologically modulated nitrification and denitrification reactions. Each is controlled by specialized groups of microbes, and adverse impacts on any of these result in a lower N availability and soil C retention.

The relatively rapid decline in the SOC pool in mine soil can be ameliorated by managing biomass productivity, the development of root formation in subsoil and the weathering of overburden (Haering et al., 1993). By promoting plant growth and productivity, SOM can accumulate in restored mine soils.

7.6   Role of root formation in carbon sequestration

Roots provide a pathway for the movement of carbon (and energy) to deeper soil horizons. Root production and turnover have a direct impact on the storage of carbon in terrestrial ecosystems.

An understanding of below-ground processes is essential to understanding their inter-relationship with above-ground processes. The transfer of carbon from roots into soil (as SOC) is poorly understood

(Izaurralde et al., 2001). Roots contribute to SOC, upon death or through exudation of organic substances during growth. These exudates, called rhizodeposits, consist of soluble compounds, secretions, lysalates and decaying fine roots and together they have the potential to store more carbon than is possible at the soil surface. The contribution made by roots is linked to their productivity and turnover, mycorrhizal colonization but also with vegetation type. Grier et al. (1981) and Santantonio and Grace (1987) reported the contribution of fine roots to SOC from 33 to 67% of the annual NPP in a forest ecosystem. Balesdent and Balabane (1996) used a δ ^{13}C technique to calculate the contribution of to SOC storage in soils in France, and found that these incorporated $57\,g\,C\,m^{-2}\,year^{-1}$; a rate 58% higher than incorporation by leaves and stalks together. These results suggest that a high production of root matter degrades more slowly, enhancing the total C retained in soil. Swinnen et al. (1995) used ^{14}C pulse labelling to study rhizodeposition of winter wheat and spring barley. They found that total rhizodeposition (i.e. $450–990\,kg\,C\,ha^{-1}\,year^{-1}$) accounted for about 7–15% of net plant assimilation and twice as much as the mass of roots left after harvest. Therefore, deep-rooted plant species have the potential to increase SOC by transferring more OM into deeper soil horizons.

7.7 Reclamation via re-vegetation to enhance carbon sequestration

The reclamation of mine soil should include the ground engineering via grading, benching and ensuring the stability of slopes. Once spoil is placed, heavy equipment should only be used for final grading. Mine spoil particles, smaller than 2 mm, are responsible for majority of a soils water- and nutrient-holding capacity in the mine soils, whereas those larger than 2 mm cannot hold enough water to sustain vigorous growth over the summer dry months. In general, 1 to 1.5 m of non-compacted (loose) soil material is required to hold enough water to sustain plants through prolonged droughts (Sheoran et al., 2010).

Compaction can be minimized during grading by using small dozers and by planting deep-rooted perennials. The planned storage and placement of topsoil over spoil at least 30–50 cm depth will significantly enhance the establishment of vegetation. The mixing of topsoil and its inoculation (e.g., N_2-fixing and P-solubilizing microbes) can also improve the quality of prepared ground. To increase seed viability, topsoil should be moved directly from the active mining area to the reclamation area, as native soil serves as an important source of gene bank for plant species suitable for re-vegetation.

Mine spoil composition also influences its C storage potential, through:

- The link between spoil type and aboveground or belowground biomass production; and
- The ability of the mine spoil to stabilize C inputs into slowly cycling OM pools

Soil acidity is a key impactor, and together with the mobile metals influences nutrient transformation and litter decomposition (Tate, 1985; Vimmerstedt et al., 1989). Both the clay content and type of clay mineral present in mine spoil (Parton et al., 1987; Motavalli et al., 1994) influence soils ability to store C in chemically recalcitrant organo-mineral complexes and stable soil aggregates (Golchin et al., 1994). The addition of organic amendments on soil bulk density (Martens and Frankenberger, 1992; Shiralipour et al., 1992) and water infiltration is more pronounced in coarse textured soils and is strongly related to increasing SOC (Khaleel et al., 1981).The engineering of soil 'chemical and 'structural' properties can both increase SOM (Tisdall and Oades, 1982) and improve soil aggregate stability (Gallardo-Lara and Nogales, 1987; Martens and Frankenberger, 1992). Changes to a spoil's physical properties, such as bulk density and water holding capacity etc. can further increase the success of managed re-vegetation (Khaleel et al., 1981; Pagliai et al., 1981; Martens and Frankenberger, 1992; Shiralipour et al., 1992; McConnell et al., 1993; Turner et al., 1994; Stockdale et al., 2001).

7.8 Ecosystem productivity and C sequestration

Several factors govern the successful re-vegetation of mine soils. The accumulation of soil C and the improvement of soil quality, depends somewhat on the ability of reclaimed soil to support and sustain vegetation. Key factors that affect the establishment of trees on reclaimed mine soils include(Torbert and Burger, 2000):

- Soil depth
- Soil toxicity
- Soil compaction; and
- Competition from weeds

Rooting depth is important to tree productivity. The soil compaction of a soil and its rock content poses physical limitations to plant growth (e.g., restricts root penetration, and impacts upon water holding capacity) within the rooting zone (Ashby et al., 1984). Andrews et al. (1992)

examined 78 reclaimed mine sites planted with white pines across a three- US state regions and observed rooting depth to be the most important factor affecting tree height. Important management practices that SOC in aggrading terrestrial ecosystems that are generally enhanced by permanent perennial vegetation, include:

- Increased input rates of OM
- Changing the decomposition rate of OM to increase litter and the light organic fraction
- Incorporation of OM deeper in to soil, either directly by enhancing below-ground inputs or indirectly by surface-mixing soil organisms (e.g. earthworms); and
- Enhanced physical properties through formation of intra-aggregate and organic-mineral complexes (Post and Kwon, 2000)

7.8.1 Topsoil

Topsoil has an important role in the succession of native species and maximizing the return from the seed resource in topsoil, is key to restoring the full suite of pre-mining species to restored land.

The use of a cover of topsoil over mine spoil provides improved growth conditions for plants. However, the stored topsoil is not always effective and a freshly borrowed soil borrowed from nearby is a more effective amendment as it supports the growth of aerobic bacteria (Ghose, 2005). Stockpiled topsoil (up to 1 m deep) supports the growth of anaerobic bacteria but not aerobic bacteria (Harris et al., 1989).

Once the soil is removed from a stockpile and reinstated, the aerobic microbial population rapidly re-establishes itself, usually at higher than the normal level (Williamson and Johnson, 1991). Nitrification restarts at higher than the normal rates (Sheoran et al., 2010); however, topsoil stored for long periods of time tends to lose its viability (Nicolau, 2002; Moreno-de las Heras et al., 2008). However, it is possible to negate this by using a 15 cm cover of 'fresh' soil to stimulates nitrification and reduce the leaching of nutrients.

In dry soil where topsoil is very thin, subsoiling is used to break up compacted soil layers, involves deep tilling of the ground (300 to 600 mm) and is different to 'topsoiling' in its depth, which usually only extends to 150 mm deep. Subsoiling can be carried out before planting to improve drainage, water infiltration, and aeration and to encourage plant root penetration and proliferation.

7.8.2 Selection of plant species

It is normal to choose drought-resistant, fast-growing and tolerant plant species to overcome nutrient deficiency in dry regions. (Figure 7.3).

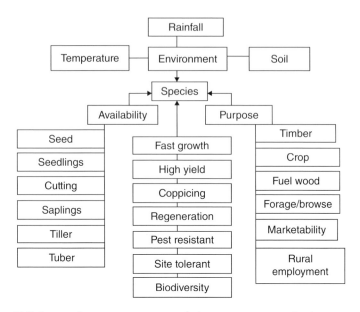

Figure 7.3 Factors determining screening of plant species in wastelands.

The use of native seed mixtures is preferable to ensure good germination of plants selected for their ability to thrive under severe conditions. Plants that tolerate difficult conditions tend to be very hardy and are unaffected by the presence of metal contaminants (Caravaca et al., 2002; Mendez and Maier, 2008) and make a significant contribution to the physical stabilisation of soil (Madejon et al., 2006).

The choosing of native flora and fauna (e.g., from a nearby the mining area) can increase the success of re-vegetation. Species with similar growth forms to the original vegetation, and that thrive in comparable soil types/conditions are the most appropriate. Another strategy is to find natural analogues for the previous indigenous species, because they are most likely to become fully established (Li et al., 2003; Chaney et al., 2007). Native legumes are the most suitable species because of their ability to contribute to soil fertility. Early successional species and cover crops can be considered to protect against soil erosion, however, plant species which rapidly generate biomass are preferred.

Grass species can be considered as a nurse crop for early vegetational purposes. Grasses have both positive and negative effects on mine soil, as although they stabilize soil, they may compete with woody species. Particularly, C4 grasses have a superior tolerance to drought, low soil nutrient levels and are 'resistant' to climatic stresses. Grass roots are fibrous and reduce erosion potential, and tend to form stabilized, moist, organic-rich soil and can compete with weedy species. An initial grass cover also encourages the development of self-sustaining plant communities (Shu et al., 2002; Singh et al., 2002; Hao et al., 2004).

Plants improve soil by increasing litter and root exudates and enhanced biological activity (Pulford and Watson, 2003; Coates, 2005; Padmavathiamma and Li, 2007; Mertens et al., 2007).

As nitrogen is a major limiting nutrient, the regular addition of nitrogenous fertilizer may be required to maintain a healthy persistent cover of vegetation (Yang et al., 2003; Song et al., 2004). Alternatively legumes and other nitrogen-fixing species improve soil fertility as the mineralization of their N-rich litter from these species allow substantial can be transferred to companion species (Zhang et al., 2001), and this and their effect on soil properties has been observed by Singh et al. (2002).

7.8.3 Amendments

7.8.3.1 *Mulching*

The application of surface mulches after seeding improves soil moisture holding capacity the availability of nutrients and moderates soil temperature. Mulches have no particular effect upon soluble salt, soil pH, or the exchange of Na or other cations.

Mulches suppress weeds and increase OM but only in the shorter-term. The application of a good sawdust, paper waste and other plant residues with high C:N/C:P ratio can contribute to OM and may even immobilize of N, P and other minerals (Singh et al., 1991a).

Straw mulch applied at the rate of $6\,Mg\,ha^{-1}$ after sowing grasses and legumes increases standing biomass. Studies have shown that a combined application of straw and wood fibre mulches greatly improves plant establishment and long-term vigour (Sheoran et al., 2010). Sawdust has shown promise to increase the survival rate of certain trees, forbs and shrubs (Uresk and Yamamoto, 1986) and it is suggested that the addition of wood chips to bare mine spoil is second only to topsoil, for establishing plant growth (Smith et al., 1985; Gitt and Dollhopf, 1991). Wood residue amendments also increase the effect of fertilizer and gypsum (Voorhees and Uresk, 1990; Sheoran et al., 2009).

In some cases, bark can significantly increase soil microbe activity raising the rate of biomass decomposition, however, bark is also known to reduce the availability of nitrate-N (Elkins et al., 1984).

The mixing of topsoil with mine soil dilutes the organic carbon present (Visser et al., 1984), sometimes to the point that the total amount of organic carbon available is a limiting factor for raising microbial metabolic activity (Williamson and Johnson, 1991). Amending soil with bark (Elkins et al., 1984) or fertilizing and planting ryegrass (Williamson and Johnson, 1991) helps bacteria (with enough organic carbon) by stimulating their metabolic activity.

7.8.3.2   *Organic amendments*

Organic amendments, such as manures and composts, provide a ready source of C and usually N to spoil. Manure from livestock provides a valuable source of SOM and nutrients, as it readily decomposes and both C and N are readily taken-up by microorganisms and ultimately, in higher plants. The application of organic amendments also improves soil mineral CEC. Leguminous plant, such as Dalbergia sissoo increases field moisture content by as much as 7%, reduces pH to ca. 5.5 and can raise organic carbon by as much as 85%. The latter is rendered by the accumulation and decomposition of leaf litter to form humus (Maiti and Ghose, 2005; Tripathi and Singh, 2008).

Recycled organics, including, biosolids, compost, biochar and manure provide the direct addition of carbon to soil. Long-term experiments in Europe have shown that soils treated with organic amendments have 20–100% more SOC than those treated with inorganic fertilizers (Lal, 2008).

7.8.3.3   *Biosolids*

Biosolids are often an excellent amendment for use with mine soils where the original topsoil contains low OM or where topsoil substitutes such as overburden material or subsoil are used. Biosolids as sewage sludge result from wastewater treatment and can be available in a number of forms, including liquid, sludge cake (dewatered and dried, heat dried, alkaline treated) or composted (Walker, 1994). The application of biosolids decreases soil bulk density, improves water-holding capacity and stabilizes pH and serve as microbial inocula. Biosolids are widely recognized as effective short-term fertilizers and a source of long-term slow-release nitrogen (Hall, 1984; Munshower, 1994; Sydnor and Redente, 2002).

Studies have reported bio-solids for improving the mine soil, including bulk density (Bendfeldt et al., 2001), acidity (Pichtel et al., 1994) resulting in higher yield compared to lime- and topsoil-amendments (Roberts et al, 1988b). Following plantation, there is a lag in the input of litter and root turnover and thus the restoration of soil carbon (Paul et al., 2002). Bio-solid addition can maintain soil carbon input even after the disturbance of establishment (Kelly, 2005).

7.8.3.4   *Microbial biomass*

Soil microbial biomass and the soil metabolic activity are sensitive indicators that respond to changes in soil environment. Soil microbial populations have metabolic versatility and are adaptable to low nutrient levels and an adverse chemical environment of mine wastes (Tripathi et al., 2014). A positive linear correlation between soil aggregate stability and microbial biomass carbon suggests the productivity of the microbial community is crucial to soil fertility (Edgerton et al., 1995). Soil microbial populations can be used to determine the stability of a restored ecosystem, as a positive correlation between plant root biomass and soil microbial biomass was observed by Tripathi et al. (2012).

Nutrient cycling is closely linked to active soil microbial biomass. Microbes process carbon, nitrogen, and phosphorus and are involved in complex interactions with soil and plants. Thus, if there is a decline in microbial populations, carbon and nitrogen cycles are disrupted. A strong relationship exists between the time of restoration and the increase in soil microbial biomass (Tripathi and Singh, 2008), as the SOM during reclamation serves as the starting point' for the development of key soil textural characteristics. The development of sustainable micro-biological processes depends on this, and the input rate and decomposition of organic matter. These in turn are partially related to soil depend upon temperature, moisture, pH and soil air content (Oades, 1988; Piao et al., 2001). The generation of plant litter and root exudates supports microbe-mediated nutrient cycling (Tripathi et al., 2009) and thus, the complex nature of competing and synergistic relationship required to sustain an ecosystem should not be underestimated.

7.9 Carbon dioxide offset from mine soils

Mining-related activities cause severe soil degradation via the loss of SOC, soil erosion and leaching of minerals and nutrients (Lal et al., 1998). Lowered SOC decreases the physical protection against soil aggregate decomposition, and a loss of 80% of the original SOC pool has been observed from scraped topsoil (Akala and Lal, 2001; Ghose, 2001). In a study assessing the shelf-life of mine soils in India, Ghose (2001) observed a 47% decrease in SOC in the first year, followed by gradual decrease until a steady-state level was achieved at 20% (of the initial concentration), at 6 years. Akala and Lal (2001) estimated more than 70% of the SOC pool is lost due to surface mining. The SOC content of reclaimed mine soils tend to be much lower than undisturbed soils and therefore, there is potential to replace C by using appropriate management practices.

The SOM increases with soil depth as OM has a density around one quarter that of the mineral soil and it aggregates soil particles. Increased OM results in lowered soil bulk density (Tunstall, 2010). Soil OM contains carbon and nitrogen which are linked to global biogeochemical cycling of carbon and nitrogen.

The mechanisms responsible for stabilizing SOC may be categorized (Christensen, 1996) as:

- Physical protection
- Biochemical recalcitrance; and
- Chemical stabilization

Chemical stabilisation involves an association between decomposable organic compounds and soil mineral components (e.g. organic C sorbed to clay surfaces). Biochemical recalcitrance is rendered by the chemical characteristics of substrate itself (e.g. lignin and their derivatives) or fungal melanin (Haider and Martin, 1981). The nature of various organo-mineral associations and their location and distribution within soil aggregates determines the extent of physical protection and chemical stabilization of SOC (Gjisman and Sanze, 1998). Clay-sized organic-mineral complexes often show greater accumulations and subsequently more rapid loss rates than in silt-sized particles, indicating a higher stability of silt–SOC (Christensen, 1996). These same mechanisms regulate the stabilisation and retention of SOC in mine soil.

Schlesinger (1990) compiled data on long-term rates of the accumulation of soil organic carbon accumulation in during the Holocene and found a slow rate of increase in carbon even after thousands of years. This long-term increase involves passive soil organic carbon, which includes charcoal and resistant compounds physically protected in organic-mineral complexes. Schlesinger (1990) documented long-term rates of carbon storage from $0.2 \, g \, C \, m^{-2} \, year^{-1}$ in some polar deserts to $C \, m^{-2} \, year^{-1}$ over all ecosystems (Post and Kwon, 2000).

Soil microbial processes regulate ecosystem net primary productivity (Cleveland et al., 2006) and most of the nutrient requirements of terrestrial plants are met via mineralised organic nutrient mediated by the microbial community (Paul and Clark, 1996). The contribution of soil microbial biomass to carbon recycling, however, is seriously underestimated and its turnover poorly understood. Microbial biomass is the most important indicator of biological soil processes, because it is 'a primary catalyst of bio-geochemical processes and the energy and nutrient reservoir' (Kutsch et al., 2010). In mine spoil, microbes sequestrate carbon because of their high tolerance towards stress conditions. Microbes have physiological mechanisms to remain active during stress conditions and to acclimate to stress by altering the allocation of from growth to survival, unless the stress is too great and irreversible (Farrar and Reboli, 1999; Suzina et al., 2004).

Though the microbial biomass is only a small fraction of SOC, it effectively mediates the transfer of SOC, low fraction and high fraction (organic-mineral) organic carbon. As a result, rates of transfer and transformation are influenced by both biologically important factors, and soil properties moisture and temperature (Post and Kwon, 2000).

7.10   Carbon accretion in revegetated mine soils

Revegetated mine spoils are a huge potential sink for CO_2. Shrestha and Lal (2009) evaluated the CO_2 offset rates from the different ways reclaimed mine spoils are used (Table 7.1). They calculated that a reclaimed forest

Table 7.1 Potential CO_2 offset land uses after reclamation.

Potential land uses		Potential CO₂ offset rate (Mg ha⁻¹ year⁻¹)	References
Forest	Biomass	6.35	Kant and Kreps (2004)
	Soil	2.28*	Sperow (2006)
		5.45*	Akala and Lal (2000)
		8.81†	Ussiri and Lal (2005)
		8.75‡	Singh et al. (2006)
	Total§	9.40	Sperow (2006)
	Biomass	4.59¶	
	Soil	1.35*	
Pasture	Soil	5.25*	Sperow (2006)
	Soil	5.39*	Akala and Lal (2000)
Cropland	Soil	3.56*	Sperow (2006)
Forest	Soil	1.20*	Present study
	Biomass	8.7	Present study
	Total§	9.36	Present study

Source: From Shrestha and Lal (2009).
* 0–30 cm depth.
† Black locust forest for 0–50 cm depth.
‡ 0–20 cm depth under 5-year-old *Albizia lebbeck* and *A. procera* plantation.
§ Total includes soil, biomass and litter.
¶ Total includes aboveground, belowground and litter mass.

mine soil could offset 30 teragrams (Tg) of CO_2 pa (based on the C seques-tration rate of 2.28 Mg ha⁻¹ year⁻¹ for forest ecosystem, taken from Sperow, 2006). They also observed that revegetated mine spoils in the United States could offset approximately 1.5 petagrams (Pg) of CO_2 produced by burning coal over 50 years.

The potential for carbon sequestration in soil depends on its current carbon level. A degraded soil with low SOC will have a greater potential to sequester carbon than a long-term managed soil (Lal, 1997; Felton et al., 2000; Subak, 2000).

Returning degraded land to perennial vegetation (pasture or trees) can substantially increase the rate of carbon sequestered (0.3–0.5 t C ha⁻¹ year⁻¹) (Potter et al., 1999; Post and Kwon, 2000). Lal (1997) estimated that the global carbon sequestration rate from restored land was 3.0 Pg year⁻¹, being 24 times greater than achieved via conservation tillage (0.125 Pg year⁻¹). By contrast, poorly managed degraded land acts as an opposite - as a source of carbon! a source of carbon.

The re-vegetation of mine soils is primarily aimed at ecosystem recov-ery. Reclaimed soils develop recognizable horizonation more quickly than otherwise, and sequester C (Thomas and Jansen, 1985). A distinct 'A' horizon (up to 15 cm thick) can develop within 5 years after reclama-tion (Thomas and Jansen, 1985; Roberts et al., 1988a, 1988b; Akala and Lal, 2000) and is distinguished from subsoil by the accumulation of SOM, loose soils due to root growth and soil structure development (Akala and Lal, 2000). In a chronosequence-focused study of mine soil reclamimed in southwest Virginia, Daniels and Amos (1981) observed that an A horizon (up to 13 cm thick) after 5 years. Thomas and Jansen (1985) investigated eight coal mine spoils and observed that the most apparent change in all sites was the development of an A horizon with darker colour, high SOM content and well-defined genetic soil structure.

In the United States, there are about 3.2 Mha of mine soil with a C sequestration potential of 0.5–1 MgC ha^{-1} year^{-1} (Lal, 2000; Office of Surface Mining (OSM), 2003), thereby potentially sequestering 1.6–3.2 TgC year^{-1} into soil and offsetting 5.8–11.7 TgCO$_2$ year^{-1} emitted by burning coal. In India, the standing biomass (i.e. above- and belowground biomass) is about 8375 Mt, of which the carbon storage would be 4178 Mt. It is estimated that from 0.85 million cubic m. biomass, 0.48 Mt of carbon are immobilised (Atkinson and Gundimeda, 2006; Tripathi et al., 2012).

Carbon uptake rates ranging from 0.2 to 2 MgC ha^{-1} year^{-1} are used to predict the likely effects of reclaiming mine land in Europe and the United States (IPCC, 2000; Nabuurs et al., 2000). Post and Kwon (2000) reported an average accumulation of 0.34 MgC ha^{-1} year^{-1} in the top 30 cm soil for a variety of tropical to temperate forest. A survey of land change data from cultivation to perennial vegetation or permanent grasslands, showed higher accumulations in permanent grassland (Post and Kwon, 2000).

In another chronosequence-related study to assess the SOC - sink capacity of reclaimed mine soil, Akala and Lal (2001, 2002) reported that the top 0–30 cm of forest and grassland can sequester 37–45 MgC ha^{-1} and 47–79 MgC ha^{-1} over 21- and 25-year periods, respectively. They observed the rate of C accumulation was higher (2–3 Mg ha^{-1} year^{-1}), initially which decreased (0.2 Mg ha^{-1} year^{-1}) after 21 years. Soils amended with topsoil reached an equilibrium SOC level in the top 15 cm earlier than soils without topsoil addition (Akala and Lal, 2002).

The time after re-vegetation has a significant effect on soil microbial biomass and the accumulation of SOC (Tripathi and Singh, 2008, 2012). Nitrogen accumulation and cycling are the most important factors in soil development. The OC content of re-vegetated spoil increases with time, due to substantial accumulation of plant, root, litter and microbial biomass (Tripathi and Singh, 2008). The enhanced recovery of mine spoil is shown by the difference between vegetative cover between 1 and 20 years (Fig. 7.4). Singh et al. (2002) observed a linear increase in SOC with time, following planting of Albizia. The uptake of carbon into re-vegetated spoil is given in Fig. 7.5.

High levels of organic matter in soil can improve its aggregation and infiltration capacities and the availability of nutrients (NRC, 2001). In forest soils, SOC content varies depending on tree type and on soil physical properties (Morris and Paul, 2003). The long-term fate of soil C depends on the residence time of SOC pools, and consumption of the aboveground biomass. However, harvesting the forest and burning the biomass returns

C directly to the atmosphere, but wood-based products continue to be a significant reserve of terrestrial C (Nabuurs et al., 2000).

Good management practices allow, tree biomass to accumulate levels as high as in natural soils. The rate of C sequestration depends on the

1 Year

4 Years

Figure 7.4 A view of revegetated mine spoils at various ages after reclamation.

8 Years

17 Years

Figure 7.4 (*Continued*)

17 Years

20 Years

Figure 7.4 *(Continued)*

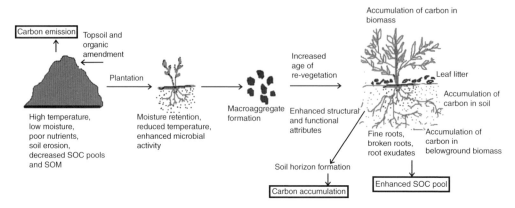

Figure 7.5 The uptake of carbon into revegetated mine spoil.

productivity of the forest or grasses. In temperate regionsed, 40% of the C taken over 25-years in a forest plantation rotation remained stored within the solid wood products (Winjum and Schroeder, 1997). On average, deciduous and pine temperate forest vegetation contain 131 and 160 $MgCha^{-1}$, respectively, with 130 $MgCha^{-1}$ stored in SOM (Houghton, 1995). Vogt et al. (1995) reported 40–62% of total ecosystem C was contained in living biomass, whereas SOM contained 33–50% of the total.

The duration of net SOC in reclaimed mine soils is uncertain. Soil organic carbon has been found to increases rapidly over the first 20 years following reclamation (Akala and Lal, 2001; Bendfeldt et al., 2001) and then accumulates at lower rates between 20 and 30 years. Model predictions based on 0–25-year chronosequence data indicate it may take 100–150 years for SOC pool to reach an equilibrium in mine soils (Akala and Lal, 2001, 2002).

7.11   Carbon sequestration in mine soil: The prospects for coal producers

Coal accounts for 24% of total energy consumed globally, making it the second largest source of primary energy. There is an estimated 11000 Gt of coal reserve globally; about 80% of those resources are found in Russia and the former Soviet Republics, the United States and China (Mitsch and Jørgensen, 2004). Global coal consumption in year 2002 was 4.8 Pg, and was projected to increase by 1.5% pa between 2008 and 2035 (EIA, 2004; IEO, 2011).

Major coal-consuming country China account in total for about 49% of the world's coal consumption (EIA, 2014).

When surface mining is used for coal extraction, large areas of land are disturbed, CO_2 is emitted from the decomposition of the OM, and fallen aboveground biomass. Global data on the area of land disturbed by surface mining activities is scarce. In India, it has been reported that for every Mt of coal extracted by surface mining, about 4 ha of land is damaged, and the coal industry renders an area of 1400 ha year^{-1} biologically unproductive (Ghose, 2001; Tripathi et al., 2014). In the United States, coal extracted from nearly 745 surface mines operating in 2011 increased 25% between 1992 and 2011 from 590 Mt to 749 Mt (National Mining Association (NMA, 2004; EIA, 2011).

In Ohio, the total area disturbed by coal mining between 1977 and 2000 was 25 918 ha (OSM, 2002). Prior to then 110 872 ha of disturbed land needed reclamation (Lal et al., 2004). In the United States, the total area disturbed by surface mining between 1977 and 2000 was nearly 617 327 ha (OSM, 2002), and 632 801 ha previously disturbed remains unreclaimed (Lal et al., 2004). Thus, the total area disturbed by coal mining in the United States was approximately 1.25 Mha. Nevertheless, based on figures from the U.S. Army Corps of Engineers for the years prior to the SMCR Act (1977), and the Office of Surface Mining upto 2008, approximately 3.4 Mha land is disturbed by surface mining (Source watch, 2011).

The SOC sequestration potential for Ohio mine soils ranges from 0.68 to 6.33 Mt C with an average of 3.50 Mt C. The SOC sequestration potential for reclaimed mine soils in the United States ranges from 6.25 to 57.8 Mt C (with an average of 32.0 Mt C, at the rate of 1.28 Mt C year^{-1}). Albeit, the United States' average C emissions from coal combustion from 1990 to 2012 declined by 10% (from 1797 Mt to 1613 Mt) (USEPA, 2004; IEA, 2014), these are still higher than the potential for SOC sequestration in mine soils (Ussiri and Lal, 2005). However, there is an additional C sequestration potential related to biomass accumulation occurring in trees, shrubs and grasslands.

7.12 Carbon sequestration activities in India

The strategy for C management in India is in accordance with the President's Global Climate Change Initiative. It also supports the goals of the International Framework on Climate Change.

The Department of Science and Technology, Ministry of Coal and Ministry of Rural Development, Govt. of India, are funding C management projects in revegetated mine wasteland. The focus is the sequestration potential of terrestrial ecosystem and how C stored in revegetated mine spoil might offset CO_2 emissions while providing additional income to landowners through the trading of C credits.

Reforestation activities by the State Forest Department's Initiative examine methods to promote post-mining land-use activities on abandoned surface coal mines. The state reforestation programme establishes forests in abandoned mine lands of Dhanbad, Bokaro, Raniganj, Singrauli (funded by the Ministry of Coal, Govt. of India) to enhance C storage.

Academic research institutes (Department of Botany, Banaras Hindu University, Varanasi; Indian Institute of Forest Research, Dehradun), national laboratories (Central Institute of Mining and Fuel Research, Dhanbad; National Environmental Engineering Institute, Nagpur), non-governmental organizations and private sectors are also actively investigating C sequestration in mine wastelands.

7.13   The carbon budget for reclaimed mine ecosystems

To identify sustainable management options and enhance carbon storage potential, a longer-term evaluation of the carbon budget is required. An examination of total carbon in different components of a revegetated site, namely, plant biomass, soil and microbial biomass after 19 years was found to be $69.21\,t\,ha^{-1}$. The total plant biomass, mine soil and soil microbial biomass carbon contributions were 64, 33 and 3%, respectively (Singh et al., 2012). There was an increase in total sequestered carbon of >700% in revegetated mine spoils, which can be translated into annual carbon sequestration potential of $3.64\,t\,ha^{-1}\,year^{-1}$.

The carbon captured from the atmosphere and stored in revegetated mine spoil is equivalent to $254\,t\,ha^{-1}$. The annual carbon budget indicated $8.40\,t\,ha^{-1}\,year^{-1}$ carbon accumulation in which: $2.14\,t\,ha^{-1}$ was allocated in aboveground biomass, $0.31\,t\,ha^{-1}$ in belowground biomass, $2.88\,t\,ha^{-1}$ in litter mass and $1.35\,t\,ha^{-1}$ in soil (Figure 7.6). This litter-mass allocation is important to SOM and microbial biomass accumulation can be predicted, as reported by several authors in Table 7.2.

An overview carbon in biomass, soil and microbial biomass is given in Fig. 7.7.

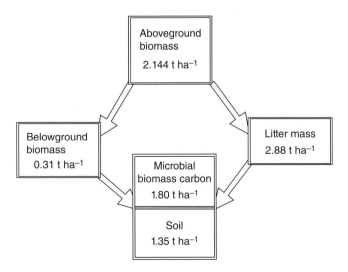

Figure 7.6 The annual carbon budget in re-vegetated coal mined wasteland.

Table 7.2 Carbon budget of grassland, forest and agricultural ecosystem.

Type of ecosystem	Location	Method of budget estimation	C budget (gCm^{-2} $year^{-1}$)	References
Grassland ecosystem				
Grassland ecosystem *Miscanthus sinensis*	Nagano, Japan	Ecological method	−100 to −56	Yazaki et al. (2004)
Pasture	New Zealand	Mass balance and modelling	−414	Tate et al. (2000)
Grassland	Cork, Ireland	Eddy covariance	+236	Leahy et al. (2004)
Grass ($200\,kg\,N\,ha^{-1}$)	Uppsala, Sweden	Ecological method	+140	Paustin et al. (1990)
Tallgrass prairie	Texas, United States	Bowen ratio/energy balance	+50 to +80	Dugas et al. (1999)
	Oklahoma, United States	Eddy covariance	−8	Suyker and Verma (2001)
	Wisconsin, United States	Difference method 2	−410 to +70	Brye et al. (2002)
Mixed-grass prairie	North Dakota, United States	Bowen ratio/energy balance (soil flux)	+31	Frank and Dugas (2001)
Moist-mixed prairie	Alberta, Canada	Eddy covariance	−18 to +21	Flanagan et al. (2002)
Meadow	Moscow, Russia	Ecological method	+387	Larionova et al. (1998)
Forest ecosystem				
Aspen–lime–birch	Moscow, Russia	Ecological method	+135	Larionova et al. (1998)
Scots pine forest, 40 years old (*Pinus sylvestris*)	Southern Finland	Eddy covariance	+228	Kolari et al. (2004)
French pine forest (*Pinus pinaster*)	Les Landes, France	Eddy covariance	−200 to −340	Kowalski et al. (2003)
Boreal and temperate forest of Ontario	Ontario, Canada	Model: CBM-CFS2	−40	Liu et al. (2002)
Ontario's forest ecosystem	Ontario, Canada	Model: CBM-CFS2	−43	Peng et al. (2006)
Indigenous forest	New Zealand	Mass balance and modelling	−136	Tate et al. (2000)
Agricultural ecosystem				
Mix agricultural crops	Denmark	Eddy covariance	−31	Soegaard et al. (2003)
Barley – no fertilizer	Uppsala, Sweden	Difference method 2	−20	Paustin et al. (1990)
−120 kg N	Uppsala, Sweden	Difference method 2	+10	Paustin et al. (1990)
Corn – continuous	Ohio, United States	Cropland ecosystem model C (CEM)	+26	Evrendilek and Wali (2004)
−Chisel ploughed, fertilized	Wisconsin, United States	Difference method 2	−90 to +590	Brye et al. (2002)

(Continued)

Table 7.2 (Continued)

Type of ecosystem	Location	Method of budget estimation	C budget (g C m^{-2} year^{-1})	References
–No till, fertilized	Wisconsin, United States	Difference method 2	–210 to +430	Brye et al. (2002)
No till corn–soybean	North Central United States	Eddy covariance	+90	Hollinger et al. (2005)
Revegetated mine wasteland	Dry tropical ecosystems, India	Ecological method	354.79	Tripathi et al. (2012)

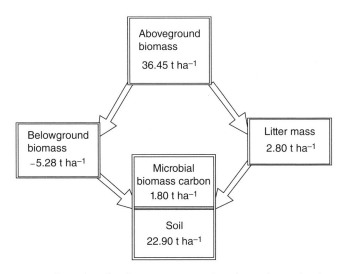

Figure 7.7 Standing state of carbon in revegetated coal mined wasteland.

7.14 Implications for management

The re-vegetation of mine spoil is an operation to return the mined land as a tool to rejuvenate and re-establish the degraded ecosystems. It should be part of an integrated programme of an effective environmental management through all phases of resource development – from exploration to construction, operation, and mine closure.

The reconstruction of mine soil quality depends on vegetation growth for improved soil physical, chemical and biological conditions. Conditions favoring the accumulation of SOM and high carbon stocks in biomass are the major outcome of the successful regeneration of mine spoil using native and late re-established vegetation.

References

Aborigines count cost of mine (2004) *BBC News/Asia-Pacific*, 25 May 2004.

Abresch, J.P., Gassner, E., von Korff, J. (2000) Naturschutz und Braunkohlesanierung. *Angew Landschaftsökol*, 27, 1–427.

Abugov, R. (1982) Species diversity and phasing of disturbance. *Ecology*, 63, 289–293.

Adriano, D.C. (1986) *Trace Elements in the Terrestrial Environment*, pp. 533. Springer, New York.

Agrawal, A., Sahu, K.K. & Pandey, B.D. (2004) Solid waste management in non-ferrous industries in India resources. *Conservation and Recycling*, 42, 99–120.

Agricultural Statistics at a Glance (2013) Department of Agriculture and Cooperation, Ministry of Agriculture, Government of India, New Delhi, India. http://eands.dacnet.nic.in/latest_2013.htm, accessed 22 June 2015.

Akala, V.A. & Lal, R. (2000) Potential of mine land reclamation for soil organic C sequestration in Ohio. *Land Degradation & Development*, 11, 289–297.

Akala, V.A. & Lal, R. (2001) Soil organic pools and sequestration rates in reclaimed mine soils in Ohio. *Journal of Environmental Quality*, 30, 2090–2104.

Akala, V. & Lal, R. (2002) Soil organic carbon sequestration rates in reclaimed mine soils. In: *Agricultural Practices and Policies for Carbon Sequestration in Soil*, (eds Kimble, J.M., Lal, R. & Follett, R.F.), pp. 297–304. Lewis Publishers, Boca Raton, FL.

Amacher, M.C., Brown, R.W., Kotuby-Amacher, J. & Willis, A. (1993) *Adding sodium hydroxide to study metal removal in a stream affected by acid mine drainage.* Research Paper INT-465, US Department of Agriculture Forest Service, Washington, DC.

Andel, J.V. & Grootjans, A.P. (2006) Concepts in restoration ecology. *Restoration Ecology*, (ed. J. van Andel & J. Aronson. Blackwell Science Ltd, Oxford.

Andrews, J.A., Torbert, J.L., Johnson, J.E. & Burger, J.L. (1992) Effects of mine soil properties on young white pine *Pinus strobus* height growth. In: *Achieving Land Use Potential Through Reclamation. Proceedings of 9th Annual Meeting*, pp. 119–129. American Society of Surface Mining and Reclamation, Duluth, MN.

Angers, D.A. & Caron, J. (1998) Plant-induced changes in soil structure: processes and feedbacks. *Biogeochemistry*, 42(1–2), 55–72.

Anon (2006) *Dirty Metals, Mining Communities and Environment*, Earthworks, Oxfam, America, Washington, DC, pp. 4.

Reclamation of Mine-Impacted Land for Ecosystem Recovery, First Edition. Nimisha Tripathi,
Raj Shekhar Singh and Colin D. Hills.
© 2016 John Wiley & Sons, Ltd. Published 2016 by John Wiley & Sons, Ltd.

Armesto, J.J. & Pickett, S.T.A. (1985) Experiments on disturbance in old-field plant communities: impact on species richness and abundance. *Ecology*, 66, 230–240.

Arnon, D.I. & Johnson, C.M. (1942) Influence of hydrogen ion concentration on the growth of higher plants under controlled conditions. *Plant Physiology*, 17, 525–539.

Aronson, J., Floret, C. & Le Floc'h, E. (1993) Restoration and rehabilitation of degraded ecosystems in arid and semi-arid regions. I. A view from the south. *Restoration Ecology*, 1, 8–17.

Ashby, C., Vogel, W.G., Kolar, C.A. & Philo, G.R. (1984) Productivity of stony soils on strip mines. In: *Erosion and Productivity of Soils Containing Rock Fragments*, (eds J.D. Nichols, P.L. Brown, & W.J. Grant, pp. 31–44. SSSA Special Publication 13, Soil Science Society of America, Madison, WI.

Aswathanarayana, U. (2003) *Mineral Resources Management and the Environment*. Balkema Publishers, Rotterdam.

Atkinson, G. & Gundimeda, H. (2006) Accounting for India's forest wealth. *Ecological Economics*, 59, 462–476.

Bai, Z.K., Zhao, J.K. & Zhu, Y.M. (1999) On the ecological rehabilitation of mined areas. *Journal of Natural Resources*, 19(2), 24–28 [in Chinese with English abstract].

Baldock, J.A. & Skjemstad, J.O. (2000) Role of the soil matrix and minerals in protecting natural organic materials against biological attack. *Organic Geochemistry*, 31, 697–710.

Balesdent, J. & Balabane, M. (1996) Major contribution of roots to soil carbon storage inferred from maize cultivated soils. *Soil Biology & Biochemistry*, 28, 1261–1263.

Banning, N.C., Gleeson, D.B., Grigg, A.H., et al. (2011) Soil microbial community successional patterns during forest ecosystem restoration. *Applied and Environmental Microbiology*, 77 (17), 6158–6164.

Bardgett, R.D. & Walker, L.R. (2004) Impact of coloniser plant species on the development of decomposer microbial communities following deglaciation. *Soil Biology & Biochemistry*, 36, 555–559.

Barley, K.P. (1954) Effects of root growth and decay on permeability of synthetic sandy loam. *Soil Science Society of America Journal*, 78, 205–211.

Baron, J.S., Poff, N.L., Angermeier, P.L., et al. (2002) Meeting ecological and societal needs for freshwater. *Ecological Applications*, 12 (5), 1247–1260.

Barrios, E., Kwesiga, F., Buresh, R.J., & Sprent, J.I. (1997) Light fraction soil OM and available nitrogen following trees and maize. *Soil Science Society of America Journal*, 61, 826–831.

Bartelmus, P. (1997) Measuring sustainability: data linkage and integration. In: *Sustainability Indicators: A Report on Indicators of Sustainable Development* (eds B. Moldan, S. Billharz & R. Matravers). Chichester, UK: John Wiley & Sons, Ltd.

Bartens, J., Day, S.D., Harris, J.R., Dove, J.E. & Wynn, T.M. (2008) Can urban tree roots improve infiltration through compacted subsoils for stormwater management? *Journal of Environmental Quality*, 37 (6), 2048–2057.

Baskin, C.C. & Baskin, J.M. (1998) *Seeds: Ecology, Biogeography and Evolution in Dormancy and Germination*, pp. 666. Academic Press, San Diego, CA.

Batjes, N.H. (1996) Total carbon and nitrogen in the soils of the world. *European Journal of Soil Science*, 47, 151–163.

Bauters, T.W.J., Steenhuis, T.S., DiCarlo, D.A., et al. (2000) Physics of water repellent soils. *Journal of Hydrology*, 231–232, 233–243.

Bazzaz, F.A. (1996) *Plants in Changing Environments. Linking Physiological, Population and Community Ecology.* Cambridge University Press, Cambridge, UK, pp. 320.

Bell, F.G. (1998) *Environmental Geology. Principles and Practice.* Cambridge University Press Cambridge, UK.

Bendfeldt, E.S., Burger, J.A., & Daniels, W.L. (2001) Quality of amended mine soils after sixteen years. *Soil Science Society of America Journal*, 65, 1736–1744.

Bhushan, C. & Hazra, M.Z. (2005) *Concrete Facts: Green Rating of Indian Cement Industry*, pp. 35. Centre for Science & Environment, New Delhi, India.

Bierman, P.M. & C.J. Rosen (2005) *Nutrient Cycling and Maintaining Soil Fertility, Fruit and Vegetable Crop Systems*, pp. 24. University of Minnesota Extension, Minneapolis.

Binelli, E.K., Gholz, H.L. & Duryea, M.L. (2008) *Plant Succession and Disturbances in the Urban Forest Ecosystem.* Chapter 4: In Restoring the Urban Forest Ecosystem. Pub.No. FOR93. IFAS, University of Florida. Florida. http://edis.ifas.ufl.edu/pdffiles/fr/fr06800.pdf. Accessed 26 June 2015.

Bierwirth, P.N. (2015) *Effects of rising carbon dioxide levels on human health via breathing toxicity - A critical issue that remains unapprehended.* ResearchGate DOI:10.13140/RG.2.1.3297.9368. (http://grapevine.com.au/~pbierwirth/co2toxicity.pdf; accessed on 19/10/2015).

Birdsey, R.A. & Heath, L.S. (1995) Carbon changes in U.S. forests. In: *Productivity of America's Forests and Climate Change* (ed. L.A. Joyce), GTR-RM-271, pp. 56–70. USDA Forest Service, Rocky Mountain Forest and Range Experiment Station, Fort Collins, CO.

Bockheim, J.G., Gennadiyev, A.N., Hammer, R.D. & Tandarich, J.P. (2005) Historical development of key concepts in pedology. *Geoderma*, 124, 23–36.

Bonan, G.B. (2008) Forests and climate change: forcings, feedbacks, and the climate benefits of forests. *Science*, 320, 1444–1449.

Borgegård, S.O. (1990) Vegetation development in abandoned gravel pits; effects of surrounding vegetation, substrate and regionality. *Journal of Vegetation Science*, 1, 675–682.

Bormann, F.H. & Likens, G.E. (1979) *Pattern and Process in a Forested Ecosystem.* Springer-Verlag, New York.

Bossuyt, H., Six, J. & Hendrix, P.F. (2005) Protection of soil carbon by micro aggregates within earthworm casts. *Soil Biology & Biochemistry*, 37, 251–258.

Bowen, G.D. & Rovira, A.D. (1999) The rhizosphere and its management to improve plant growth. *Advances in Agronomy*, 66, 1–102.

BP p.l.c. (2012) *BP Statistical Review of World Energy.* BP p.l.c., London. http://www.bp.com/liveassets/bp_internet/globalbp/globalbp_uk_english/reports_and_publications/statistical_energy_review_2011/STAGING/local_assets/pdf/statistical_review_of_world_energy_full_report_2012.pdf, accessed 11 March 2013.

BP Statistical Energy Survey (2007) www.ficci.com/spdocument/20317/Mining-Industry.pdf, accessed 27 June 2015.

Bradshaw, A.D. (1983) The reconstruction of ecosystems. Presidential address to the British Ecological Society. *Journal of Applied Ecology*, 20, 1–17.

Bradshaw, A.D. (1987a) Restoration: the acid test for ecology. In *Restoration Ecology: A Synthetic Approach to Ecological Research*, (eds W.R. Jordan, M.E. Gilpin & J.D. Aber), pp. 23–29. Cambridge University Press, Cambridge, UK.

Bradshaw, A.D. (1987b) What do we mean by restoration? In *Restoration Ecology and Sustainable Development*, (eds K.M. Urbanska, N.R. Webb & P.J. Edwards), pp. 8–16. Cambridge University Press, Cambridge, UK.

Bradshaw, A.D. (1996) Underlying principles of restoration. *Canadian Journal of Fisheries and Aquatic Sciences*, 53(1), 3–9.

Bradshaw, A.D. (1997) What do we mean by restoration? In *Restoration Ecology and Sustainable Development*, (eds K.M. Urbanska, N.R. Webb & P.J. Edwards), pp. 8. Cambridge University Press, Cambridge, UK.

Brady, N.C. & Weil, R.R. (2002) *The Nature and Properties of Soils*, 13th edn, pp. 960. Pearson Education Inc., Upper Saddle River, NJ.

Brant, J.B. & Myrold, D.D. (2006) Root controls on soil microbial community structure in forest soils. *Oecologia*, 148(4), 650–659.

Brauman, A., Bignell, D.E., & Tayasu, I. (2000) Soil-feeding termites: biology, microbial associations and digestive mechanisms. In *Termites: Evolution, Sociality, Symbioses, Ecology* (eds T. Abe, D.E. Bignell, & M. Higashi), pp. 233–259. Kluwer Academic, Dordrecht, The Netherlands.

BRGM (2001) *Management of Mining, Quarrying and Ore-Processing Waste in the European Union*, pp. 79. BRGM, Orléans, France.

British Geological Survey Commissioned Report (2006) *Mineral Planning Factsheet: Limestone*. http://www.bgs.ac.uk/downloads/start.cfm?id=1361, accessed 4 June 2015.

British Geological Survey Commissioned Report (2010) *Mineral Planning Factsheet: Coal*. http://www.bgs.ac.uk/downloads/start.cfm?id=1354, accessed 4 June 2015.

British Geological Survey Commissioned Report (2011) *Mineral Planning Factsheet: Ball Clay*. https://www.bgs.ac.uk/downloads/start.cfm?id=1348, accessed 4 June 2015.

Bronick, C.J. & Lal. R. (2005) Soil structure and management: a review. *Geoderma*, 124(1–2), 3–22.

Brooker, R.W. & Callaghan, T.V. (1998) The balance between positive and negative plant interactions and its relationship to environmental gradients: a model. *Oikos*, 81, 196–207.

Brown, S. (1997) *Estimating Biomass and Biomass Change of Tropical Forests: A Primer, FAO Forestry Paper 134*. Food and Agriculture Organization of the United Nations, Rome.

Brown, C. (2013) *North Dakota went boom. New York Times Magazine*, 3 February.

Brown, R.W. & Amacher, M.C. (1999) Selecting plant species for ecological restoration: a perspective for land managers. In: *Revegetation with Native Species: Proceedings, 1997 Society for Ecological Restoration Annual Meeting; 1997 November 12–15; Fort Lauderdale, FL* (ed. L.K. Holzworth, comp. R.W. Brown). pp. 1–16, RMRSP-8. U.S. Department of Agriculture, Forest Service, Rocky Mountain Research Station, Fort Collins, CO.

Brown, R.W. & Chambers, J.C. (1990) Reclamation practices in high mountain ecosystems. In: *Proceedings-Symposium on Whitebark Pine Ecosystems: Ecology and Management of a High-Mountain Resource; 1989 March 29–31; Bozeman, MT*.

pp. 329–334, Gen. Tech. Rep. INT-GTR-270. U.S. Department of Agriculture, Forest Service, Intermountain Research Station, Ogden, UT.

Brown, S. & Subler, S. (2007) Generating carbon credits through mine site restoration with organic amendments. In: *Mine Closure 2007: Proceedings of the Second International Seminar on Mine Closure, 16–19 October 2007, Santiago, Chile* (eds A. Fourie, M. Tibbett & J. Wiertz), pp. 459–464. Salviat Impresores, Santiago, Chile.

Brudvig, L.A. (2011) The restoration of biodiversity: where has research been and where does it need to go? *American Journal of Botany*, 98(3), 549–558.

Brunckhorst, D.J. (2010) Using context in novel community-based natural resource management: landscapes of property, policy and place. *Environmental Conservation*, 37 (1), 16–22.

Brundtland Report WCED (1987) Report of the World Commission on Environment and Development Our Common Future. United Nations, New York City.

Burger, J.A. (1999) *Powell River project – restoring the value of forests on reclaimed mined land*. www.cses.vt.edu/PRP/ and Virginia Cooperative Extension (www.ext.vt.edu), accessed 12 June 2015.

Brye, K.R., Gower, S.T., Norman, J.M. & Bundy, L.G. (2002) Carbon budgets for a prairie and agro-ecosystems: effects of land use and inter annual variability. *Ecological Applications*, 12, 962–979.

Bureau of Land Management (2001) http://www.blm.gov/wo/st/en/prog/energy/coal_and_non-energy.print.html, accessed 7 August 2011.

Buyanovsky, G.A., Aslam, M. & Wagner, G.H. (1994) Carbon turnover in soil physical fractions. *Soil Science Society of America Journal*, 58, 1167–1173.

Cairns J., Jr. (1986) Restoration, reclamation, and regeneration of degraded or destroyed ecosystems. *Conservation Biology*, (ed. M.E. Soule), pp. 465 Sinauer Publishers, Ann Arbor, MI.

Cairns, J. (1995) Restoration ecology: protecting our national and global life support systems. In: *Rehabilitating Damaged Ecosystems* (ed. J. Cairns), pp. 1–12. Lewis Publishers, Boca Raton, FL.

Callaway, R.M. (1995) Positive interactions among plants. *Botanical Review*, 61, 306–349.

Cambardella, C.A. & Elliot, E.T. (1993) Carbon and nitrogen distribution in aggregates from cultivated and native grassland soils. *Soil Science Society of America Journal*, 56, 1071–1076.

CAMMA (2001) *Conference of Mining Ministers of the Americas*. Memorandum of Understanding: General Non-Binding Principles for mine Closure in the Americas. Cited in Research Mine Closure policy. MMSD, 2002 No.44. IIED and WBCSD. England. (http://pubs.iied.org/pdfs/G00541.pdf)

Campusano, R.F. & Patricio, C. (eds) (2001) *Cierre de Faenas Mineras en Chile: Propuesta de Legislación, Institucionalidad y Opciones técnicas*. IDRC, MPRI, COCHILCO, Santiago.

Canham, C.D., Berkowitz, A.R., Kelly, V.R., Lovett, G.M., Ollinger, S.V. & Schnurr, J. (1996) Biomass allocation and multiple resource limitation in tree seedlings. *Canadian Journal of Forest Research*, 26, 1521–1530.

Caravaca, F., Barea, J.M., Figueroa, D. & Roldán, A. (2002) Assessing the effectiveness of mycorrhizal inoculation and soil compost addition for enhancing reafforestation with *Olea europaea* subsp. *sylvestris* through changes in soil biological and physical parameters. *Applied Soil Ecology*, 20, 107–118.

Censere (2012) Appraise, Assist, Advice China Coal Industry. May http://www.censere.com/index.php/en/articles-publications/research-reports/product/down load/file_id-5/lang-en-GB

Censere Report (2012) *China Coal Industry*, pp. 52. www.censere.com/index.php/en/articles...reports/product/...5/lang-en-GB.

Central Pollution Control Board (CPCB) (2000) Report on management of municipal solid wastes, Delhi, India. *Economic Survey of India, 2008–09*, Ministry of Finance, New Delhi, India.

Cernea, M.M. (2000) Risks, safeguards, and reconstruction: a model for population displacement and resettlement. In: *Risks and Reconstruction: Experiences of Resettlers and Refugees* (eds M.M. Cernea & C. McDowell), pp. 11–55. World Bank, Washington, DC.

Chamber of Mines of South Africa (2012) *Facts and Figures, 2011: Johannesburg, South Africa*, pp. 41. Chamber of Mines of South Africa, Johannesburg.

Chaney, R.L., Angle, J.S., Broadhurst, C.L., Peters, C.A., Tappero, R.V. & Donald, L.S. (2007) Improved understanding of hyperaccumulation yields commercial phytoextraction and phytomining technologies. *Journal of Environmental Quality*, 36, 1429–1443.

Chapin, F.S., III, Matson, P.A. & Mooney, H.A. (2002) *Principles of Terrestrial Ecosystem Ecology*. New York: Springer. pp. 281–304.

Chapin, F.S., III, Matson, P.A. & Vitousek, P.M. (2011) Temporal dynamics. *Principles of Terrestrial Ecosystem Ecology*, Springer, New York. pp. 339–367.

Chevallier, T., Blanchart, E., Albrecht, A. & Feller, C. (2004) The physical protection of soil organic carbon in aggregates: a mechanism of carbon storage in a vertisol under pasture and market gardening (Martinique, West Indies). *Agriculture, Ecosystems and Environment*, 103, 375–387.

Chhabra, A., Palria, S. & Dadhwal, V.K. (2003) Soil organic carbon pool in Indian forests. *Forest Ecology and Management*, 173, 187–199.

Chong, S.K., Becker, M.A., Moore, S.M. & Weaver, G.T. (1986) Characterization of reclaimed mined land with and without topsoil. *Journal of Environmental Quality*, 15 (2), 157–160, doi:10.2134/jeq1986.00472425001500020014x.

Christensen, B.T. (1996) Carbon in primary and secondary organo-chemical complexes. In: *Structure and Organic Matter in Agricultural Soils* (eds M.R. Carter & B.A. Stewart), pp. 97–165. CRC Lewis, Boca Raton, FL.

Ciolkosz, E.J., Cronce, R.C., Cunningham, R.L. & Peterson, G.W. (1985) Characteristics, genesis, and classification of Pennsylvania minesoils. *Soil Science*, 139, 232–238.

Clark, J.C. (1996) *Law and policy concerning social and cultural issues in mining. Proceedings of 5th Asia/Pacific Mining Conference and Exhibition*, Jakarta, Indonesia, 1996.

Clark, D.S. (1999) The many meanings of the rule of law. In: *Law, Capitalism and Power in Asia* (ed. K. Jayasuriya). Routledge, London.

Clements, F.E. (1916) *Plant Succession: An Analysis of the Development of Vegetation*, Carnegie Institute of Washington, Washington, DC. pp 512.

Clements, F.E. (1928) *Plant Succession and Indicators: A Definitive Edition of Plant Succession and Plant Indicators*, The H.W. Wilson Company, New York.

Cleveland, C.C., Reed, S.C. & Townsend, A.R. (2006) Nutrient regulation of organic matter decomposition in a tropical rain forest. *Ecology*, 87, 492–503.

Coal Mining in India-Overview (2005) www.mbendi.com/indy/ming/coal/ as/in/p0005.htm, accessed 4 June 2015.

Coal Statistics (2013) *Energy Trends: September 2013. Special Feature Article – Coal in 2013*. Department of Energy and Climate Change, Government of UK, London. https://www.gov.uk/government/uploads/system/uploads/ attachment_data/file/357548/Coal_2013.pdf, accessed 12 June 2015.

Coates, W. (2005) Tree species selection for a mine tailings bioremediation project in Peru. *Biomass Bioenergy*, 28 (4), 418–423.

Connell, J.H. (1978) Diversity in tropical rainforests and coral reefs. *Science*, 199, 1302–1310.

Connell, J.H. & Slatyer, R.O. (1977) Mechanisms of succession in natural communities and their role in community stability and organization. *American Naturalist*, 111, 1119–1144.

Daft, M.J. & Hacskaylo, E. (1976) Arbuscular mycorrhizas in the anthracite and bituminous coal wastes of Pennsylvania. *Journal of Applied Ecology*, 13, 523–530.

Davies, R., Hodgkinson, R., Younger, A. & Chapman, R. (1995) Nitrogen loss from a soil restored after surface mining. *Journal of Environmental Quality*, 24, 1215–1222.

Daniels, W.L. & Amos, D.F. (1981) Mapping, characterization and genesis of minesoils on a reclamation research area in Wise County, Virginia. In: *Proceedings, Symposium on Surface Mining, Hydrology, Sedimentation, and Reclamation* (ed. D.H. Graves), pp. 261–265. University of Kentucky, Lexington.

Daniels, W.L. & Zipper, C.E. (1999) *Creation and management of productive mine soils*. Publication 460-131, Virginia Cooperative Extension, Virginia Polytechnic Institute and State University, Blacksburg. http://www.vt.edu/ pubs/mines/460-121.pdf; accessed on 12 August 2015.

Danielson, L. & Nixon, M. (2000) Current regulatory approaches to mine closure in the United States. In: *Environmental Policy in Mining* (eds A. Warhurst & L. Noronha) pp. 311–350. Lewis Publishers, Boca Raton, FL.

Day, S.D., Wiseman, P.E., Dickinson, S.B. & Harris, J.R. (2010). Tree Root Ecology in the Urban Environment and Implications for a Sustainable Rhizosphere. *Arboriculture & Urban Forestry*, 36(5): 193–205.

Dean, C., Roxburgh, S.H., Harper R.J., Eldridge, D.J., Watson, I.W., & Wardell-Johnson, G.W. (2012) Accounting for space and time in soil carbon dynamics in timbered rangelands. *Ecological Engineering*, 38, 51–64.

Deelwal, K., Dharavath, K. & Kulshreshtha, M. (2014) Evaluation of characteristic properties of red mud for possible use as a geotechnical material in civil construction. *International Journal of Advances in Engineering & Technology*, 7 (3), 1053–1059.

Denman, K.L., Brasseur, G., Chidthaisong, A., (2007) Couplings between changes in the climate system and biogeochemistry. In: *Climate Change 2007: The Physical Science Basis. Contribution of Working Group I to the Fourth Assessment Report of the Intergovernmental Panel on Climate Change*, (eds S. Solomon, D. Qin, M. Manning, Z. Chen, M. Marquis, K.B. Averyt, M. Tignor & H.L. Miller), pp. 501–587. Cambridge University Press, Cambridge, UK.

Dent, D. (2007) Land. In: *Global Environmental Outlook GEO4* (ed. UNEP), pp. 81–114. United Nations Environment Programme, Nairobi.

Department of Energy, South Africa (2012) http://www.energy.gov.za/files/links_frame.html, accessed 27 June 2015.

Dexter, A.R. (1988) Advances in the characterization of soil structure. *Soil and Tillage Research*, 11, 199–238.

Dirección General de Asuntos Ambientales (1999) *Incluye la modificación sugún D.S. N° 041-2001-EM (2001-07-21)*. www.actualidaambiental.pe/…/Roi-de-las-autoridades….

Dixon, R.K., Brown, S., Solomon, R.A., Trexler, M.C., & Wisniewski, J. (1994) Carbon pools and flux of global forest ecosystems. *Science*, 263, 185–190.

Dodd, C.K., & Smith L.L. (2003) Habitat destruction and alteration: historical trends and future prospects for amphibians. In: *Amphibian Conservation*, (eds R.D. Semlitsch). pp. 94–112. Smithsonian Institution, Washington, DC.

Doha Climate Change Conference (2012) *United Nation Convention on Climate Change. Earth Negotiations Bulletin*. 26 November-8 December, Doha, Qatar. http://unfccc.int/meetings/doha_nov_2012/meeting/6815.php; accessed on 27 June, 2015.

Dokuchaev, V.V. (1893) Abridged historical account and critical examination of the principal soil classifications existing. *Transactions of the Petersburg Society of Naturalists*, 1, 64–67.

Dollhopf, D.J. & Baumm, B.J. (1981) *Bentonite mine land reclamation in the Northern Great Plains*. Report no. 179, Montana Agricultural Experiment Station, Reclamation Research Program, Montana State University, Bozeman, MT.

Downing, T.E. (2002) Avoiding new poverty: mining-induced displacement and resettlement. *Mining, Minerals and Sustainable Development, No. 58*, IIED, London.

Draft IBM Manual on Appraisal of Mining Plan (2013) *Presented at Workshop on Mining Plan Guidelines and RQP Examination, Organized by Training Centre IBM, Nagpur, 15–16 April 2013*.

Dragovich, D. & Patterson, J. (1995) Condition of rehabilitated coal mines in the Hunter Valley, Australia. *Land Degradation & Rehabilitation*, 6, 29–39.

Dugas, W.A., Heuer, M.L. & Mayeux, H.S. (1999) Carbon dioxide fluxes over bermuda grass, native prairie, and sorghum. *Agricultural and Forest Meteorology*, 93, 121–139.

Dulal, H.B., Brodnig, G. & Shah, K.U. (2011) Capital assets and institutional constraints to implementation of greenhouse gas mitigation options in agriculture. *Mitigation and Adaptation Strategies for Global Change*, 16 (1), 1–23.

Earth Negotiations Bulletin (2012) Vol. 12, No. 539, 18 May 2012.

Earth's CO_2 home page. http://co$_2$now.org/Current-CO$_2$/CO$_2$-Now/global-co$_2$-board.html;accessed 12 December, 2012.

Edgerton, D.L., Harris, J.A., Birch, P. & Bullock, P. (1995) Linear relationship between aggregate stability and microbial biomass in three restored soils. *Soil Biology & Biochemistry*, 27, 1499–1501.

Egan, D. & Howell, E.A. (eds.) (2001) *The Historical Ecology Handbook: A Restorationist's Guide to Reference Ecosystems*, Island Press, Washington, DC.

Egler, F.E. (1954) Vegetation science concepts. I. Initial floristic composition, a factor in old field vegetation development. *Vegetatio*, 4, 412–417.

EIA (Energy Information Administration)/Annual Energy Outlook (2004) *With Projections to 2025*. U.S. Energy Information Administration, Washington, DC. www.eia.doe.gov/oiaf/aeo/.

EIA (2011) *International Energy Statistics 2011*. U.S. Energy Information Administration, Washington, DC. http://tonto.eia.doe.gov/cfapps/ipdbproject/ IEDIndex3.cfm, accessed 11 June 2015.

EIA (2012) Global CO2 emissions from fossil fuel combustion were taken from Energy Information Administration International Energy Statistics http:// tonto.eia.doe.gov/cfapps/ipdbproject/IEDIndex3.cfm; accessed on 30th October, 2015.

EIA (2014) China produces and consumes almost as much coal as the rest of the world combined http://www.eia.gov/todayinenergy/detail.cfm?id=16271; accessed on 1st November, 2015.

EIA (2015) Report on Energy and Climate Change. https://www.iea.org/ publications/freepublications/publication/WEO2015SpecialReporton EnergyandClimateChange.pdf; accessed on 30th October, 2015.

Elkins, N.Z., Parker, L.W., Aldon, E. & Whitford, W.G. (1984) Responses of soil biota to organic amendments in stripmine spoil in northwestern New Mexico. *Journal of Environmental Quality*, 13, 215–219.

Elliott, H.A., Linn, J.H. & Shields, G.A. (1989) Role of Fe in extractive decontamination of Pb-polluted soils. *Hazardous Waste and Hazardous Materials*, 6, 223–228.

Embacher, A. (2000) *Wasser-und Stoffhaushalt einer Eichenchronosequenz auf kohle-und schwefelhaltigen Kippsubstraten der Niederlausitz.* (Cottbuser Schriften zu Bodenschutz und Rekultivierung, vol. 10). Brandenburgische Technische Universität, Cottbus, Germany

Energy Information Administration (2006) *Country Analysis Briefs: China.* http:// eia.gov/countries/analysisbriefs/China/china.pdf; accessed 11 July 2015.

Environmental Performance Report of SAIL (2005) MINENVIS, a newsletter of the ENVIS center on environmental problems in mining area, No. 44, Indian School of Mines, p. 9, March 2005. http://s3.amazonaws.com/ zanran_storage/www.ismenvis.nic.in/ContentPages/2469071535.pdf; accessed 11 July 2015.

Epron, D., Bahn, M., Derrien, D., et al. (2012) Pulse-labelling trees to study carbon allocation dynamics: a review of methods, current knowledge and future prospects. *Tree Physiology*, 32, 776–798.

Evrendilek, F. & Wali, M.K. (2004) Changing global climate: historical carbon and nitrogen budgets and projected responses of Ohio's cropland ecosystems. *Ecosystems*, 7, 381–392.

Falk, D.A., Palmer, M.A. & Zeadler, J.B. (2006) *Foundations of Restoration Ecology. Society for Ecological Restoration International*, pp. 355. Island Press, Washington, DC.

Farrar, W. & Reboli, A. (1999) The genus *Bacillus*: medical. In: *The Prokaryotes: An Evolving Electronic Resource for the Microbiological Community* (ed. M. Dworkin), 3rd edn., release 3.0. Springer-Verlag, New York.

Farrar, J., Hawes, M., Jones, D. & Lindow, S. (2003) How roots control the flux of carbon to the rhizosphere. *Ecology*, 84, 827–837.

Fehrenbacher, D.J., Jansen, I.J. & Fehrenbacher, J.B. (1982) Corn root development in constructed soils on surface mined land in Western Illinois. *Soil Science Society of America Journal*, 46, 353–359.

Felton, W., Schwenke, G., Martin, R. & Fisher, J. (2000) Changes in carbon in summer rainfall cropping systems. In: *Management Options for Carbon*

Sequestration in Forest, Agricultural and Rangeland Ecosystems, (eds R. Keenan, A.L. Bugg & H. Ainslie), pp. 32–37. CRC for Greenhouse Accounting, Canberra.

Fenner, M. (1987) Seed characteristics in relation to succession. In: *Colonization, Succession and Stability* (eds. J. Gray, M.J. Crawley & P.J. Edwards). Blackwell Scientific Publications, London.

Filip, Z. (2002) International approach to assessing soil quality by ecologically-related biological parameters. *Agriculture, Ecosystems and Environment*, 88, 169–174.

Filley, T.R., Boutton, T.W., Liao, J.D., Astrow, J.D. & Gamblin, D.E. (2008) Chemical changes to non-aggregated particulate soil organic matter following grassland-to-woodland transition in subtropical savanna. *Journal of Geophysical Research: Biogeosciences*, 113, G03009. doi:10.1029/2007JG000564.

Finch, W.I., Butler, A.P., Armstrong Jr., F.C. & Weissenborn, A.E. (1973) Nuclear fuels. In: *United States Mineral Resources*, US Geological Survey, Professional Paper 820, pp. 458.

Flanagan, L.B., Wever, L.A. & Carlson, P.J. (2002) Seasonal and interannual variation in carbon dioxide exchange and carbon balance in northern temperate grassland. *Global Change Biology*, 8, 599–615.

Fogel, R. & Hunt, G. (1983) Contribution of mycorrhizae and soil fungi to nutrient cycling in a Douglas-fir ecosystem. *Canadian Journal of Forest Research*, 13(2), 219–232.

Forman, R.T.T. & Godron, M. (1981) Patches and structural components for a landscape ecology. *BioScience*, 31, 733–740.

Fortin, P. (1992) Recent trends in mineral development laws. In: *Mineral Industry Taxation Policies for Asia and the Pacific, United Nations Development Programme*, p. 23. United Nations, New York.

Frank, A.B. & Dugas, W.A. (2001) Carbon dioxide fluxes over a northern, semiarid, mixedgrass prairie. *Agricultural and Forest Meteorology*, 108, 317–326.

Gallaher, W.J. & Lynn, S. (1989) *A Review of Hardrock Mine Reclamation Practices as Background for Proposed Nevada Legislation*. Public Resource Associates, Reno, NV.

Gallardo-Lara, F. & Nogales, R. (1987) Effect of the application of town refuse compost on the soil-plant system: a review. *Biological Wastes*, 19, 35–62.

Ghose, M.K. (2001) Management of topsoil for geo-environmental reclamation of coal mining areas. *Environmental Geology*, 40, 1405–1410.

Ghose, M.K. (2005) Soil conservation for rehabilitation and revegetation of mine-degraded land. *TIDEE–TERI Information Digest on Energy and Environment*, 4(2), 137–150.

Gibbs, R.J. & Reid, J.B. (1988) A conceptual model of changes in soil structure under different cropping systems. *Advances in Soil Science*, 8,123–149.

Gitt, M.J. & Dollhopf, D.J. (1991) Coal waste reclamation using automated weathering to predict lime requirement. *Journal of Environmental Quality*, 20, 285–288.

Gjisman, A.J. & Sanze, I.I. (1998) Soil organic matter pools in volcanic ash soil under fallow or cultivation with applied chicken manure. *European Journal of Soil Science*, 49, 427–436.

Glinski J. & Lipiec, J. (1990) *Soil Physical Conditions and Plant Roots*, CRC Press, Boca Raton, FL.

Goebel, M.O., Bachmann, J., Woche, S.K. & Fischer, W.R. (2005) Soil wettability, aggregate stability, and the decomposition of soil organic matter. *Geoderma*, 128, 80–93.

Golchin, A., Oades, J.M., Skjemstad, J.O. & Clarke, P. (1994) Study of free and occluded particulate organic matter in soils by solid state 13C CP/MAS NMR spectroscopy and scanning electron microscopy. *Australian Journal of Soil Research*, 32, 285–309.

Golldack, J., Münzenberger, B. & Hüttl, R.F. (2000) Mykorrhizierung der kiefer (*Pinus sylvestris* L.) auf forstlichrekultivierten kippenstandorten des Lausitzer Braunkohlereviers. In: *Rekultivierung in Bergbaufolgelandschaften* (eds. G. Broll, W. Dunger, B. Keplin & W. Topp), pp. 131–146. Springer, Berlin.

Gonzalez-Perez, M.A. & Leonard, L. (eds) (2013) *International Business, Sustainable and Corporate Social Responsibility*. Advances in Sustainability and Environmental Justice, Vol. 11. Emerald, Bingley. https://books.google.co.uk/books?isbn=1781906254, accessed 16 June 2015.

Gopalakrishnan, T. (2006) http://www.academia.edu/5782806/Paper, accessed 27 June 2015.

Goss, M.J. (1991) Consequences of the activity of roots on soil. In: *Plant Root Growth: An Ecological Perspective* (ed D. Atkinson). Blackwell Scientific Publications, Oxford.

Gould, A.B., Hendrix, J.W. & Ferriss, R.S. (1996) Relationship of mycorrhizal activity to time following reclamation of surface mine land in western Kentucky. I. Propagule and spore population densities. *Canadian Journal of Botany*, 74, 247–261.

Grace, P.R. & Basso, B. (2012) Offsetting greenhouse gas emissions through biological carbon sequestration in North Eastern Australia. *Agricultural Systems*, 105, 1–6.

Grayston, S.J., Vaughan, D. & Jones, D. (1997) Rhizosphere carbon flow in trees, in comparison with annual plants: the importance of root exudation and its impact on microbial activity and nutrient availability. *Applied Soil Ecology*, 5(1), 29–56.

Greenfield, J.C. (1989) *ASTAG Tech. Papers*. The World Bank, Washington, DC.

Greenland, D.J., Wild, A. & Adams, A. (1992) Organic matter dynamics in soils of the tropics – from myth to complex reality. In: Lal, R., Smith, P.A., editors. *Myths and Science of Soils of the Tropics*, pp. 17–39. Soil Science Society of America, Madison, WI.

Greenpeace International Report (2010) *Mining Impacts*. http://www.greenpeace.org/international/en/campaigns/climate-change/coal/Mining-impacts/, accessed 27 June 2015.

Grier, C.C., Vogt, K.A., Reyes, M.R. & Edmonds, R.L. (1981) Biomass distribution and above- and below-ground production in young and mature *Abies anabilis* zone ecosystems of the Washington Cascades. *Canadian Journal of Forest Research*, 11, 155–167.

Griffiths, E. & Jones, D. (1965) Microbial aspects of soil structure, I: relationships between organic amendments, microbial colonization and changes in aggregate stability. *Plant and Soil*, 23, 17–33.

Grubb, P.J. (1987) Some generalizing ideas about colonization and succession in green plants and fungi. In: *Colonization, Succession and Stability*, (eds A.J. Gray, M.J. Crawley & P.J. Edwards), pp. 81–102. Blackwell Scientific Publications, Oxford.

Gupta, T.N. (1998) *Building Materials in India: 50 Years, a Commemorative Volume*. New Delhi, India: Building Materials Technology Promotion Council, Government of India.

Haering, K.C., Daniels, W.L. & Roberts, J.A. (1993) Changes in mine soil properties resulting from overburden weathering. *Journal of Environmental Quality*, 22, 194–200.

Haider, K. & Martin, J.P. (1981) Decomposition in soil of specifically 14C carbon isotope labeled model and cornstalk lignins and coniferyl alcohol over two years as influenced by drying, re-wetting and addition of an available carbon substrate. *Soil Biology & Biochemistry*, 13, 447–450.

Hall, J.E. (1984) The cumulative and residual effects of sewage sludge nitrogen on crop growth. In: *Long-Term Effect of Sewage Sludge and Farm Slurries Applications*, (ed J.H. Williams, G. Guidi, P. L'Hermite), pp. 73–83. Elsevier Applied Science Publisher, London.

Han, K.H., Ha, S.G. & Jang, B.C. (2010) Aggregate stability and soil carbon storage as affected by different land use practices. In: *Proceedings of International Workshop on Evaluation and Sustainable Management of Soil Carbon Sequestration in Asian Countries, Bogor, Indonesia, 28–29 September 2010*. http://www.niaes.affrc.go.jp/marco/bogor2010/proceedings/00.pdf, accessed 26 June 2015.

Hao, X.Z., Zhou, D.M., Wang, Y.J. & Chen, H.M. (2004) Study of rye grass in copper mine tailing treated with peat and chemical fertilizer. *Acta Pedologica Sinica*, 41(4), 645–648.

Harrington, C.A. (1999) Forests planted for ecosystem restoration or conservation, *New Forest*, 17, 175.

Harris, J.A. & Diggelen, R.V. (2006) Ecological restoration as a project for global society. In: *Restoration Ecology* (ed J. van Andel & J. Aronson), pp. 3–15. Blackwell Publishing, Oxford.

Harris, J.P., Birch, P. & Short, K.C. (1989) Changes in the microbial community and physio-chemical characteristics of top soils stockpiled during opencast mining. *Soil Use and Management* 5, 161–168.

Harris, J.A., Birch, P. & Palmer, J. (1996) *Land Restoration and Reclamation: Principles and Practice*. Longman Higher Education, Harlow.

Hartman, H. (1987) *SME Mining Engineering Handbook*. pp. 1277. Society of Mining Engineering, Littleton, CO.

Heinsdorf, D.L. (1992) *Untersuchungen zur Düngerbedürftigkeit von Forstkulturen auf kipprohböden der Niederlausitz*. TU Dresden, Tharandt, Germany.

Heinselman M.L. (1973) Fire in the virgin forests of the Boundary Waters Canoe Area, Minnesota. *Quaternary Research*, 3, 329–382.

Heras, M.M.L. (2009) Development of soil physical structure and biological functionality in mining spoils affected by soil erosion in a Mediterranean-Continental environment. *Geoderma*, 149, 249–256.

Higgs, E.S. (1997) What is good ecological restoration? *Conservation Biology*, 11, 338–348.

Hindustan Copper Ltd. *Opportunities in Indian Copper Mining Industry*. http://mines.nic.in/, accessed 24 December 2014.

Hollinger, S.E., Bernacchi, C.J. & Meyers, T.P. (2005) Carbon budget of mature no-till ecosystem in North Central Region of the United States. *Agricultural and Forest Meteorology*, 130, 59–69.

Horn, R. & Dexter, A.R. (1989) Dynamics of soil aggregation in irrigated desert loess. *Soil and Tillage Research*, 13(3), 253–266.

Houghton, R.A. (1995) Changes in the storage of terrestrial carbon storage since 1850. In: *Soils and Global Change*, (eds R. Lal, J.M. Kimble, E. Levine & B.A. Stewart, pp. 45–65. CRC Lewis, Boca Raton, FL.

Houghton, R.A., Boone, R.D., Melillo, J.M., Palm, C.A., Woodwell, G.M. & Myers, N. (1985) Net flux of carbon dioxide from terrestrial tropical forests in 1980, *Nature*, 316, 617–620.

http://blogs.scientificamerican.com/observations/2013/05/09/400-ppm-carbon-dioxide-in-the-atmosphere-reaches-prehistoric-levels/, accessed 19 March 2015.

http://envfor.nic.in, accessed 12 June 2015.

http://india.indymedia.org/en/2002/12/2456.shtml. Mine & Mineral waste in India; accessed 8th August, 2015.

http://moef.nic.in/downloads/home/home-SoE-Report-2009.pdf, accessed 12 June 2015.

http://www.scclmines.com/scclnew/company-about-us.asp; accessed 26th July, 2015.

http://timesofindia.indiatimes.com/india/New-land-acquisition-law-comes-into-force/articleshow/28204302.cms, accessed 12 January 2014.

http://www.cci.in/pdfs/surveys-reports/Mineral-and-Mining-Industry-in-India.pdf, accessed 12 June 2015.

http://www.cseindia.org/node/386, accessed 12 June 2015.

http://www.cseindia.org/programme/industry/mining/political_minerals_mapdescription.htm, accessed 12 June 2015.

http://www.eldoradochemical.com/fertiliz1.htm, accessed 12 June 2015.

http://www.globalrestorationnetwork.org/restoration/methods-techniques/, accessed 12 June 2015.

http://www.iea.org/textbase/nppdf/free/2000/coalinindia2002.pdf, XXXX.

http://www.indiaenergyportal.org/subthemes_link.php?text=nuclear&themeid=12, accessed 27 June 2015.

http://www.moef.gov.in, accessed 12 June 2015.

http://www.tifac.org.in/index.Php?Option=comcontentSolidWaste GenerationandUtilizationinCalcareousStoneIndustry, 27 accessed 2015.

http://www.waste-management-world.com/articles/2003/07/an-overview-of-the-global-waste-to-energy-industry.html, 27 accessed 2015.

Hu, Z., Caudle, R.D. & Chong, S.K. (1992) Evaluation of firm land reclamation effectiveness based on reclaimed mine properties. *International Journal of Surface Mining, Reclamation and Environment*, 6, 129–135.

Huston, M.A. (1979) A general hypothesis of species diversity. *The American Naturalist*, 113, 81–101.

IBM Report, Government of India. http://ibm.nic.in/writereaddata/files/1020 2014112509msmpmar14_07_Highlights_Eng.pdf, accessed 26 June 2015.

IEA (2012) Global carbon-dioxide emissions increase by 1.0 Gt in 2011 to record high. http://www.iea.org/newsroomandevents/news/2012/may/; accessed 2nd August, 2015.

IEA (2013) *CO₂ Emissions From Fuel Combustion Highlights*, 2013 edn. International Energy Agency, Paris. http://www.iea.org/publications/freepublications/publication/CO2EmissionsFromFuelCombustionHighlights2013.pdf, accessed 23 March 2015.

IEA (2014). CO_2 Emissions from fuel combustion Highlights, 2014 edn. international Energy Agency. https://www.iea.org/publications/freepublications/publication/CO_2EmissionsFromFuelCombustionHighlights2014.pdf; accessed on 26th October, 2014.

IEO (2011). EIA-International Energy Outlook, US DOE, Sept. 2011, Washington DC.

Indian Minerals Yearbook (2011) Part I, 50th edn. Indian Mineral Industry and National Economy, Government of India, Ministry of Mines, Indian Bureau of Mines, Nagpur, India. www.ibm.gov.in, accessed 12 June 2015.

Indian Minerals Yearbook (2012a) 3rd edition.

Indian Minerals Yearbook (2012b) 5th edition.

Indorante, S.J. & Jansen, I.J. (1981) Soil variability on surface-mined and undisturbed land in southern Illinois. *Soil Science Society of America Journal*, 45, 564–568.

Intarapravich, D. & Clark, A.L. (1994) Performances guarantee schemes in the minerals industry for sustainable development. *Resource Policy*, 20, 59–69.

IPCC (Intergovernmental Panel in Climate Change) (2000) *Land Use, Land Use Change and Forestry*. In: *A Special Report of the IPCC* (eds R.T. Watson, I.R. Noble, B. Bolin, N.H. Ravindranath, D.J. Verardo & D.J. Dokken), pp. 377. Cambridge University Press, Cambridge, UK.

IPCC (Intergovernmental Panel in Climate Change) (2001) *Climate Change 2001: The Scientific Basis. Contribution of Working Group I to the Third Assessment Report of the Intergovernmental Panel on Climate Change*, Cambridge University Press, Cambridge, UK.

IPCC (2014) Summary for Policymakers. In: Climate Change 2014: Mitigation of Climate Change. Contribution of Working Group III to the Fifth Assessment Report of the Intergovernmental Panel on Climate Change [Edenhofer, O., R. Pichs-Madruga, Y. Sokona, E. Farahani, S. Kadner, K. Seyboth, A. Adler, I. Baum, S. Brunner, P. Eickemeier, B. Kriemann, J. Savolainen, S. Schlömer, C. von Stechow, T. Zwickel and J.C. Minx (eds.)]. Cambridge University Press, Cambridge, United Kingdom and New York, NY, USA.

IPCC (Intergovernmental Panel in Clame Change) (2015) *Climate Change 2014 Synthesis Report*. ISBN 978-92-9169-143-2

International Energy Outlook (2009) Coal. Chapter 4. Energy Information Administration. https://books.google.co.in/books?id=NkkpFMAmTqMC&pg=PA49&dq=International+Energy+Outlook+(2009)+Chapter+4,+coal,+energy+information+administration; accessed 15th July, 2015.

Irshad, M. (2013) Mining, river pollution and disaster risk reduction: an institutional analysis. *Environment and Ecology Research*, 1(3), 142–149.

Izaurralde, R.C., Rosenberg, N.J. & Lal, R. (2001) Mitigation of climatic change by soil carbon sequestration: issues of science, monitoring, and degraded lands. *Advances in Agronomy*, 70, 1–75.

Jansen, W. (2012) *Commodity lead-platinum KPMG in South Africa. Quarterly Commodity Insights Bulletin.* https://www.kpmg.com/ZA/en/IssuesAndInsights/ArticlesPublications/Quarterly-Commodity-Insights/Documents/QCIB_Platinum.pdf; accessed 12th June, 2015.

Jaques. L. & Huleatt, M. (2009). Australian mineral exploration peaks. AusGeo News. March, 93, Australian Govt.Geoscience Australia. http://www.ga.gov.au/ausgeonews/ausgeonews200903/mineral.jsp.

Jastrow, J.D., Miller, R.M. & Lussenhop, J. (1998) Contributions of interacting biological mechanisms to soil aggregate stabilization in restored prairie. *Soil Biology & Biochemistry*, 30(7), 905–916.

Jenny, H. (1941) *Factors of Soil Formation: A System of Quantitative Pedology.* McGraw-Hill/Dover, Mineola, NY.

Jenny, H. (1980) *Soil Genesis with Ecological Perspectives.* Springer-Verlag, New York.

Jepma, C.J. & Munasinghe, M. (1998) *Climate Change Policy: Facts, Issues and Analyses*, Cambridge University Press, Cambridge, UK.

Jha, A.K. & Singh, J.S. (1992) Rehabilitation of mine spoils. In: *Restoration of Degraded Lands: Concepts and Strategies*, pp. 211–253.

Johnson, M.G. (1995) The role of soil management in sequestering soil carbon. In: *Soil Management and Greenhouse Effect*, (eds R. Lal, J. Kimble, E. Levine & B.A. Stewart), pp. 351–363. Lewis Publishers, Boca Raton, FL.

Johnston, R.S., Brown, R.W. & Craven, J. (1975) Acid mine rehabilitation problems at high elevations. In: *Watershed Management Symposium.* pp. 66–79. Logan, UT: American Society of Civil Engineers, Irrigation and Drainage Division.

Johnson, F.L., Gibson, D.J. & Risser, P.G. (1982) Revegetation of unreclaimed coal strip-mines in Oklahoma. *Journal of Applied Ecology*, 19, 453–463.

Jones, D.L., Hodge, A. & Kuzyakov, Y. (2004) Plant and mycorrhizal regulation of rhizodeposition. *New Phytologist*, 163(3), 459–480.

Jordan, W.R., Gilpin, M.E. & Aber, J.D. (eds) (1990) *Restoration Ecology: A Synthetic Approach to Ecological Research*, Cambridge University Press, Cambridge, UK.

Kaewkrom, P., Kaewkla, N., Thummikkapong, S. & Punsang, S. (2011) Evaluation of carbon storage in soil and plant biomass of primary and secondary mixed deciduous forests in the lower northern part of Thailand. *African Journal of Environmental Science and Technology*, 51, 8–14.

Kant, Z. & Kreps, B. (2004) *Carbon sequestration and reforestation of mined lands in the Clinch and Powell River Valleys.* Topical report (reporting period: 30 May 2002–30 May 2004), DOE award number: DE-FC-26-01NT41151, The Nature Conservancy, Arlington, VA.

Katzur, J. & Hanschke, L. (1990) Pflanzenerträge auf meliorierten schwefelhaltigen Kippböden und die bodenkundlichen Zielgrößen der landwirtschaftlichen Rekultivierung. *Archiv für Acker- und Pflanzenbau und Bodenkunde*, 34, 35–43.

Kaye, J.P. & Hart, S.C. (1997) Competition for nitrogen between plants and soil microorganisms. *Trends in Ecology & Evolution*, 12, 139–143.

Keeling, C.D. & Whorf, T.P. (2004) *In Trends: A Compendium of Data on Global Change.* Carbon Dioxide Information Analysis Center, Oak Ridge National Laboratory, U.S. Department of Energy, Oak Ridge, TN. http://cdiac.ornl.gov/trends/co2/sio-mlo.html, accessed 27 June 2015.

Keever, C. (1979) Mechanisms of succession on old fields of Lancaster County, Pennsylvania. *Bulletin of the Torrey Botanical Club*, 106, 299–308.

Kelly, G.L. (2005) *Efficacy of municipal organic and green wastes in mine site rehabilitation. 10th European Biosolids and Biowaste Conference*, Wakefield, 13–16 November 2005.

Kenk, G. & Guehne, S. (2001) Management of transformation in Central Europe. *Forest Ecology and Management*, 151, 107.

Kennedy, B.A. (ed.) (1990) Surface Mining. 2nd Edition. Littleton Co. Society for Mining, Metallurgy, and Exploration. Remove comma after Metallurgy. 1194pp.

Kenney, R. (2007) *Surface Mining Control and Reclamation Act of 1977, United States*. Retrieved from http://www.eoearth.org/view/article/156351, accessed 27 June 2015.

Kemper, W.D., Rosenau, R.C. (1984) Soil cohesion as affected by time and water content. *Soil Sci. Soc. Am. J*, 48, 1001–1006.

Kimetu, J.M., Lehmann, J., Kinyangi, J.M., Cheng, C.H., Thies, J., Mugendi, D.N. & Pelle, A. (2009) Soil organic C stabilization and thresholds in C saturation. *Soil Biology & Biochemistry*, 41, 2100–2104.

Khaleel, R., Reddy, K.R. & Overcash, M.R. (1981) Changes in soil physical properties due to organic waste applications: a review. *Journal of Environmental Quality*, 10, 133–141.

Khoshoo, T.N. (2008) *Environmental Concerns and Strategies*, 3rd revised and enlarged edn. http://www.kaveribooks.com/index.php?p=sr&Uc= 9788131303733, accessed 27 June 2015.

Kolari, P., Pumpanen, J., Rannik, U., Ilvesniemi, H., Hari, P. & Berninger, F. (2004) Carbon balance of different aged Scots pine forests in Southern Finland. *Global Change Biology*, 10, 1106–1119.

Kolk, A. & Bungart, R. (2000) Bodenmikrobiologische Untersuchungen an forstlich rekultivierten kippenflächen im Lausitzer Braunkohlenrevier. In: *Rekultivierbodenökologische Prozesse und Standortenwicklung* (eds. G. Broll, W. Dunger, B. Keplin & W. Topp), Geowissenschaften und Umwelt, vol. 4, Springer, Berlin.

Körner, C. (2000) Biosphere responses to CO_2-enrichment. *Ecological Applications*, 10, 1590–1619.

Kowalski, S., Sartore, M., Burlett, R., Berbigier, P. & Loustau, D. (2003) The annual carbon budget of a French pine forest (*Pinus pinaster*) following harvest. *Global Change Biology*, 9, 1051–1065.

Krishan, G., Srivastav, S.K., Kumar, S., Saha, S.K. & Dadhwal, V.K. (2009) Quantifying the underestimation of soil organic carbon by the Walkley and Black technique – examples from Himalayan and Central Indian soils. *Current Science*, 96(8), 1133–1136.

Krull, E., Baldock, J., Skjemstad, J. (2001) Soil texture effects on decomposition and soil carbon storage. In *NEE Workshop Proceedings, 18–20 April 2001*. CRC for Greenhouse Accounting, CSIRO Land and Water, Glen Osmond, Australia.

Kumar, R., Pandey, S. & Pandey, A. (2006) Plant roots and carbon sequestration. *Current Science*, 91(7), 885–890.

Kumar, S., Maiti, S.K. Chaudhuri, S. & Ghosh, P. (2015) Potential Soil Organic Carbon Sequestration in Reclaimed Afforested Post-Mining Sites of Raniganj Coalfield, India. National Climate Science Conference, Bangalore (http://www.researchgate.net/publication/279314539; accessed on 16/10/2015).

Kutsch, W.L., Bahn, M. & Heinemeyer, A. (2010) Measuring soil microbial parameters relevant for soil carbon fluxes. *Soil Carbon Dynamics: An Integrated and Methodology* (eds W.L. Kutsch, M. Iahn & A. Heinemeyer). Cambridge University Press, Cambridge, UK.

Lafond, J., Angers, D.A. & Laverdiere, M.R. (1992) Compression characteristics of a clay soil as influenced by crops and sampling dates. *Soil and Tillage Research*, 22(3–4), 233–241.

Lal, R. (1997) Residue management, conservation tillage and soil restoration for mitigating greenhouse effect by CO_2 enrichment. *Soil and Tillage Research*, 43, 81–107.

Lal, R. (2000) Restorative effects of *Mucuna utilis* on soil organic C pool of a severely degraded Alfisol in western Nigeria. In: *Global Climate Change and Tropical Ecosystems*, (eds R. Lal, J.M. Kimble, B.A. Stewart), pp. 147–165. CRC Press, Boca Raton, FL.

Lal, R. (2004a) Agricultural activities and the global carbon cycle. *Nutrient Cycling in Agroecosystems*, 70, 103–116.

Lal, R. (2004b) Soil carbon sequestration impacts on climate change and food security. *Science*, 304, 1623–1627.

Lal, R. (2008) Sequestration of atmospheric CO_2 into global carbon pool. *Energy & Environmental Science*, 1, 86–100.

Lal, R. & Bruce, J.P. (1999) The potential of world cropland soils to sequester C and mitigate the greenhouse effect. *Environmental Science & Policy* 2:177–185.

Lal, R., Kimble, J. & Stewart, B.A. (1995) World soils as a source or sink for radio-actively active gases. In: *Soil Management and the Greenhouse Effect* (eds R. Lal, J. Kimble, E. Levine & B.A. Stewart), pp. 1–7, CRC/Lewis Publishers, Boca Raton, FL.

Lal, R., Kimble, J. and Follet, R. (1998) Land use and soil C pools in terrestrial ecosystems. In: *Management of Carbon Sequestration in Soil* (eds. R. Lal, J. Kimble, R. Follet & B.A. Stewart), pp. 1–10. Lewis Publishers, Boca Raton, FL.

Lal, R., Sobecki, T.M., Iivari, T. & Kimble, J.M. (2004) *Soil Degradation in the United States: Extent, Severity, and Trends*, pp. 163–171. Lewis Publishers, Boca Raton, FL.

Lamparter, A., Deurer, M. Bachmann, J. & Duijnisveld, W.H.M. (2006) Effect of subcritical hydrophobicity in a sandy soil on water infiltration and mobile water content. *Journal of Plant Nutrition and Soil Science*, 169, 38–46.

Landres, P.B., Morgan, P. & Swanson, F.J. (1999) Overview of the use of natural variability concepts in managing ecological systems. *Ecological Applications*, 9, 1179–1188.

Larionova, A.A., Yermolayev, A.M., Blagodatsky, S.A., Rozanov, I.V. & Orlinsky, D.B. (1998) Soil respiration and carbon balance of gray forest soils as affected by land use. *Biology and Fertility of Soils*, 27, 251–257.

Lasat, M.M. (2000) Phytoextraction of metals from contaminated soil: a review of plant/soil/metal interaction and assessment of pertinent agronomic issues. *Journal of Hazardous Substance Research*, 2, 1–23.

Lasat, M.M., Baker A.J.M. & Kochian, L.V. (1998) Altered Zn compartmentation in the root symplasm and stimulated Zn absorption into the leaf as mechanisms involved in Zn hyper accumulation in *Thlaspi caerulescens*. *Plant Physiology*, 118, 875–883.

Laves, D., Franko, U. & Thum, J. (1993) Umsatzverhalten fossiler organischer Substanzen. *Archiv für Acker- und Pflanzenbau und Bodenkunde*, 37, 211–219.

Lawrey, J.D. (1977) The relative decomposition potential of habitats variously affected by surface coal mining. *Canadian Journal of Botany*, 5, 1544–1552.

Leahy, P., Kiely, G. & Scanlon, T.M. (2004) Managed grasslands: a greenhouse gas sink or source? *Geophysical Research Letters*, 31, L20507, doi:10.1029/2004GL021161.

Ledin, M. & Pedersen, K. (1996) The environmental impact of mine wastes – roles of microorganisms and their significance in treatment of mine wastes. *Earth-Science Reviews*, 41, (1–2), 67–108.

Lemmnitz, C., Kuhnert, M., Bens, O., Guntner, A., Merz, B. & Hüttl, R.F. (2008) Spatial and temporal variations of actual soil water repellency and their influence on surface runoff. *Hydrological Processes*, 22, 1976–1984.

Li, M.S. (2006) Ecological restoration of mineland with particular reference to the metalliferous mine wasteland in China: a review of research and practice. *Science of the Total Environment*, 357, 38–53.

Li, Y.M., Chaney, R.L., Brewer, E.P., Angle, J.S., and Nelkin, J. (2003) Phytoextraction of nickel and cobalt by hyperaccumulator *Alyssum* species grown on nickel-contaminated soils. *Environmental Science & Technology*, 37, 1463–1468.

Liao, J.D., Boutton, T.W. and Jastrow, J.D. (2006) Organic matter turnover in soil physical fractions following woody plant invasion of grassland: evidence from natural 13C and 15N. *Soil Biology & Biochemistry*, 38, 3197–3210.

Lindemann, W.C., Lindsey, D.L. & Fresquez, P.R. (1984) Amendment of mine spoils to increase the number and activity of microorganisms. *Soil Science Society of America Journal*, 48, 574–578.

Liptzin, D. & Silver, W. (2009) Effects of carbon additions on iron reduction and phosphorus availability in a humid tropical forest soil. *Soil Biology & Biochemistry*, 41, 1696–1702.

Liu, G.H. & Shu, H.L. (2003) Research progress of ecological restoration in mine spoils. *Jiangxi Forestry Science and Technology*, 2, 21–25 [in Chinese].

Liu, J.X., Peng, C.H., Apps, M., Dang, Q.L., Banfield, E. & Kurz, W. (2002) Historic carbon budgets of Ontario's forest ecosystems. *Forest Ecology and Management*, 169, 103–114.

Liu, J., Liu, S. & Loveland, T.R. (2006) Temporal evolution of carbon budgets of the Appalachian forests in the U.S. from 1972 to 2000. *Forest Ecology and Management*, 222, 191–201.

Lorenz, K. & Lal, R. (2010) Effect of disturbance, succession and management on carbon sequestration. In: *Carbon Sequestration in Forest Ecosystems*, Springer, Dordrecht, The Netherlands.

Lorenz, K., Lal, R., Preston, C.M., Nierop, K.G.J. (2007) Strengthening the soil organic carbon pool by increasing contributions from recalcitrant aliphatic biomacromolecules. *Geoderma*, 142, 1–10.

Lynch, J.M. & Bragg, E. (1985) Microorganisms and soil aggregate stability. *Australian Journal of Soil Research*, 27, 411–423.

McClaugherty, C.A., Aber, J.D. & Melillo, J.M. (1982) The role of fine roots in the organic matter and nitrogen budgets of two forested ecosystems. *Ecology*, 63, 1481–1490.

McConnell, D.B., Shiralipour, A. & Smith, W. H. (1993) Compost application improves soil properties. *Biocycle*, 34, 61–63.

McCook, L.J. (1994) Understanding ecological community succession: causal models and theories, a review. *Vegetatio*, 110, 115–147.

MacDonald, G.M., Bennett, K.D., Jackson, S.T., Parducci, L., Smith, F.A., Smol, J.P., & Willis, K.J. (2008) Impacts of climate change on species, populations and communities: palaeobiogeographical insights and frontiers. *Progress in Physical Geography*, 32(2), 139–172.

MacMahon, J.A. & Jordan, W. (1994) Ecological restoration. In: *Principles of Conservation Biology* (eds G.K. Meffe & C.R. Carroll), pp. 409–438. Sinauer Associates, Inc. Sunderland, MA.

McSweetney, K. & Jansen, I. J. (1984) Soil structure and associated rooting behavior in minesoils. *Soil Science Society of America Journal*, 48, 607–612.

Madejon, E., de Mora, A.P., Felipe, E., Burgos, P. & Cabrera, F. (2006) Soil amendments reduce trace element solubility in a contaminated soil and allow regrowth of natural vegetation. *Environmental Pollution*, 139, 40–52.

Magnuson, J.J., Regier, H.A., Christien, W.J. & Sonzogi, W.C. (1980) To rehabilitate and restore great lakes ecosystems. In *The Recovery Process in Damaged Ecosystems* (ed. J. Cairns), pp. 95–112. Ann Arbor Science, Ann Arbor, MI.

Maiti, S.K. & Ghose, M.K. (2005) Ecological restoration of acidic coal mine overburden dumps – an Indian case study. *Land Contamination and Reclamation*, 134, 361–369.

Maiti, S.K. & Saxena, N.C. (1998) Biological reclamation of coal mine spoils without topsoil: an amendment study with domestic raw sewage and grass-legumes mixture. *International Journal of Surface Mining, Reclamation and Environment*, 12, 87–90.

Marinissen, J.C.Y. & Dexter, A.R. (1990) Mechanisms of stabilization of earthworm casts and artificial casts. *Biology and Fertility of Soils*, 9, 163–167.

Martens, D.A. & Frankenberger, W.T., Jr. (1992) Modification of infiltration rates in an organic amended irrigated soil. *Agronomy Journal*, 84, 707–717.

Mays, D. A., Sistani, K. R. & Soileau, J. M. (2000) Lime and fertilizer needs for land reclamation. In: *Reclamation of Drastically Disturbed Lands*, (eds R. I. Barnhisel, R. G. Darmody & W. L. Daniels), pp. 217–271. SSSA, Madison, IL.

Meharg, A.A. & Killham, K. (1991) A novel method of quantifying root exudation in the presence of soil microflora. *Plant and Soil*, 133, 111–116.

Mench, M.J., Didier, V.L., Loffer, M., Gomez, A. & Masson, P. (1994) A mimicked in situ remediation study of metal contaminated soils with emphasis on cadmium and lead. *Journal of Environmental Quality*, 23, 58–63.

Mendez, M.O. & Maier, R.M. (2008) Phytostabilization of mine tailings in arid and semiarid environments – an emerging remediation technology. *Environmental Health Perspectives*, 116(3), 278–283.

Méndez-Ortiz, B.A., Carrillo-Chávez, A. & Monroy-Fernández, M.G. (2007) Acid rock drainage and metal leaching from mine waste material (tailings) of a Pb-Zn-Ag skarn deposit: environmental assessment through static and kinetic laboratory tests. *Journal of Geological Sciences*, 24(2), 161–169.

Mertens, J., Van Nevel, L., De Schrijver, A., Piesschaert, F., Oosterbean, A., Tack, F. M. G. & Verheyen, K. (2007) Tree species effect on the redistribution of soil metals. *Environmental Pollution*, 1492, 173–181.

Michanek, G. (2008) *The law on mining and environmental protection in Sweden Interactions between the Mining Act and the Environmental Code*. Seminar in Rovaniemi, 25–26 September 2008. http://arcticcentre.ulapland.fi/docs/ NIEM_mining_Michanek_paper.pdf, accessed 27 June 2015.

Miles, J. & Walton, D.W.H. (1993) *Primary Succession on Land*. Blackwell, Oxford.

Mining, Minerals and Sustainable Development Report (2002) *Research on Mine Closure Policy*. Cochilco, Chilean Copper Commission, No. 44. http://pubs.iied.org/pdfs/G00541.pdf, accessed 12 June 2015.

Ministry of Environment and Forests (Department of Environment, Forests and Wildlife) (2008) No. 3A/86-FP.

Ministry of Mines (2012) *Annual Report, 2011–12*, pp. 300. Ministry of Mines, New Delhi, India.

Ministry of Mines, Government of India (2014) http://mines.nic.in/, accessed 12 June 2015.

Mitchell, A.R., Ellsworth, T.R. & Meek, B.D. (1995) Effect of root systems on preferential flow in swelling soil. *Communications in Soil Science and Plant Analysis*, 26, 2655–2666.

Mitsch,W.J. & Jørgensen, S.E. (2004) *Ecological Engineering and Ecosystem*, John Wiley, New York.

Moll, E. & Moll, G. (1994) *Struik Pocket Guide – Common Trees of Southern Africa*, Struik Publishers (Pty), Cape Town, South Africa.

Moreno-de las Heras, M., Nicolau, J.M. & Espigares, M.T. (2008) Vegetation succession in reclaimed coal-mining sloped in a Mediterranean-dry environment. *Ecological Engineering*, 34, 168–178.

Morris, S.J. & Paul, E.A. (2003) Forest soil ecology and soil organic carbon. In: *The Potential of U.S. Forest Soils to Sequester Carbon and Mitigate the Greenhouse Effect* (eds J.M. Kimble, L.S. Heath, R.A. Birdsey & R. Lal), pp. 109–125. Lewis Publishers, Boca Raton, FL.

Motavalli, P.P., Palm, C.A., Parton, W.J., Elliot, E.T. & Frey, S.D. (1994) Comparisons of laboratory and modelling simulation methods for estimating soil carbon pools in tropical forest soils. *Soil Biology and Biochemistry*, 26, 935–944.

Moura-Costa, P. (1996) Tropical forestry practices for carbon sequestration: a review and case study from Southeast Asia. *Ambio*, 25, 279–283.

Munshower, F.F. (1994) *Practical Handbook of Disturbed Land Revegetation*, pp. 288. CRC Press, Boca Raton, FL.

Musgrove, S. (1991) An assessment of the efficiency of remedial treatment of metal polluted soil. In: *Proceedings of the International Conference on Land Reclamation, University of Wales*. Elsevier, Essex, UK.

Nabuurs, G.J., Dolman, A.J., Verkaik, E., et al. (2000) Article 3.3 and 3.4 of the Kyoto protocol: consequences for industrialized countries' commitment, the monitoring needs and possible side effects. *Environmental Science & Policy*, 3, 123–134.

National Academy of Science (1974) *A Nationwide System for Animal Health Surveillance*. Prepared by a Special Panel of the Committee on Animal Health, Agricultural Board, National Research Council, pp. 56. National Academy of Science, Washington, DC.

National Academy of Sciences (1974) *Rehabilitation of Western Coal Lands*, Ballinger Press Cambridge, MA.

National Mining Association (NMA) (2004) *Facts About Coal*, NMA, Washington, DC.

Nehdi, M. & Tariq, A. (2007) Stabilization of sulphidic mine tailings for prevention of metal release and acid drainage using cementations materials: a review. *Journal of Environmental Engineering and Science*, 6(4), 423–436.

Nicolau, J.M. (2002) Runoff generation and routing in a Mediterranean-continental environment: the Teruel coalfield, Spain. *Hydrological Processes*, 16, 631–647.

Noble, I.R. & Slatyer, R.O. (1980) The use of vital attributes to predict successional changes in plant communities subject to recurrent disturbances. *Vegetatio*, 43, 5–21.

Norland, M.R. (2000) Use of mulches and soil stabilizers for land reclamation. In: *Reclamation of Drastically Disturbed Lands*, (eds R.I. Barnhisel, R.G. Darmody, & W.L. Daniels), pp. 645–666. American Society of Agronomy, Crop Science Society of America, Soil Science Society of America, Madison, WI.

NRC (2001) *Climate Change Science: An Analysis of Some Key Questions*. A report produced by the Committee on the Science of Climate Change, US National Research Council (US NRC), National Academy Press, Washington, DC.

Nyland, R.D. (2003) Even-to uneven-aged: the challenges of conversion. *Forest Ecology and Management*, 172, 291.

Oades, J. M. (1988) The retention of organic matter in soils. *Biogeochemistry*, 5, 35–70.

Odum, E.P. (1959) *Fundamentals of Ecology*, 2nd edn, pp. 546. W.B. Saunders Company, Philadelphia, PA.

Odum, E. P. (1969) The strategy of ecosystem development. *Science*, 164, 262–270.

Odum, E.P. (1971) *Fundamentals of Ecology*, 3rd edn. W.B. Saunders Company, Philadelphia, PA.

OECD (2014) Recent Developments in Steel Making Raw Material Markets. OECD/South Africa Workshop on Steel Making Raw Materials, DSTI, Cape Town, 11 December http://www.oecd.org/sti/ind/Raw%20material%20 background%20note%20final%20(revised).pdf; accessed 19th July, 2015.

Office of Surface Mining (OSM) (2002) *Annual Report*. Office of Surface Mining, U.S. Department of Interior, Washington, DC.

Office of Surface Mining (OSM) (2003) *Annual Report*. Office of Surface Mining, U.S. Department of Interior, Washington, DC.

Owen, J.R. & Kemp, D. (2015) Mining-induced displacement and resettlement: a critical appraisal. *Journal of Cleaner Production*, 87, 478–488.

Pacala S. & Socolow, R. (2004) Stabilisation wedges: solving the climate problem for the next 50 years with current technology. *Science*, 305, 968–972.

Padmavathiamma, P.K. & Li, L.Y. (2007) Phytoremediation technology: hyperaccumulation metals in plants. *Water Air Soil Pollution*, 184(1–4), 105–126.

Pagliai, M., Guidi, G., La Marca, M., Giachetti, M. & Lucamante, G. (1981) Effects of sewage sludges and composts on soil porosity and aggregation. *Journal of Environmental Quality*, 10, 556–561.

Palmer, M. A. & Bernhardt, E. S. (2006) Scientific pathways to effective river restoration. *Water Resources Research*, 42(3), W03507.

Palumbo, A.V., McCarthy, J.F., Amonette, J.E., Fisher, L.S., Wullschleger, S.D. & Daniels, W.L. (2004) Prospects for enhancing carbon sequestration and reclamation of degraded lands with fossil fuel combustion by-products. *Advances in Environmental Research*, 8, 425–438.

Papp, J.F. (2011) *Chromium: U.S. Geological Survey mineral commodity summaries*, pp. 42–43. http://minerals.usgs.gov/minerals/pubs/commodity/ chromium/mcs-2011-chrom.pdf, accessed 27 June 2015.

Papp, J.F. (2013) *Mineral Commodity Summaries*. United States Geological Survey, Reston, VA.

Pappu, A., Saxena, M. & Asolekar, S.R. (2007) *Solid wastes generation in India and their recycling potential in building materials.* http://dspace.library.iitb.ac.in/jspui/bitstream/10054/1649/1/5691.pdf, accessed 11 June 2015.

Paramguru, R. K., Rath, P. C. & Misra, V. N. (2005) Trends in red mud utilization – a review. *Mineral Processing and Extractive Metallurgy Review*, 26(1), 1–29.

Park, J.H., Panneerselvam, P., Lamb, D., Choppala, G. & Bolan, N. (2012) Role of organic amendments on enhanced bioremediation of heavy metal(loid) contaminated soils. *Journal of Hazardous Materials*, 185, 549–574.

Parsons, A. & Kilani, J. (2000) *Financial provision for mine closure: lessons from South Africa.* Chamber of Mines of South Africa. Presented in Vancouver 2000.

Parton, W. J., Schimel, D.S., Cole, C.V. & Ojima D.S. (1987) Analysis of factors controlling soil organic matter levels in Great Plains grasslands. *Soil Science Society of America Journal*, 51, 1173–1179.

Paul, E.A. & Clark, F.E. (1996) *Soil Microbiology and Biochemistry*, 2nd edn. Academic Press, San Diego, CA, 368 pp.

Paul, K.I., Polglase, P.J., Nyakuengama, J.G. & Khanna, P.K. (2002) Change in soil carbon following afforestation. *Forest Ecology and Management*, 168, 241–257.

Paustin, K., Andren, O., Clarholm, M., et al. (1990) Carbon and nitrogen budgets of four agro-ecosystems with annual and perennial crops with and without fertilization. *Journal of Applied Ecology*, 1990, 27, 60–84.

Pearce, A. & Armfield, C. (1998) *General history of Welsh mining.* http://www.users.globalnet.co.uk/~lizcolin/w_gen.htm, accessed 11 June 2015.

Pedroli, B., de Blust, G., van Loog, K. & van Rooij, S. (2002) Setting targets in strategies for river restoration. *Landscape Ecology*, 17, 5–18.

Peng, X., Horn, R., Peth, S. & Smucker, A. (2006) Quantification of soil shrinkage in 2D by digital image processing of soil surface. *Soil and Tillage Research*, 91, 173–180.

Piao, H.C., Liu, G.S., Wu, Y.Y. & Xu, W.B. (2001) Relationships of soil microbial biomass carbon and organic carbon with environmental parameters in mountainous soils of southwest China. *Biology and Fertility of Soils*, 33, 347–350.

Pichtel, J.R., Dick, W.A. & Sutton, P. (1994) Comparison of amendments and management practices for long-term reclamation of abandoned mine lands. *Journal of Environmental Quality*, 23, 766–772.

Pickett, S.T.A. & Thompson, J.N. (1978) Patch dynamics and the design of nature reserves. *Biological Conservation*, 13, 27–37.

Pickett, S.T.A. & White, P.S. (1985) Natural disturbance and patch dynamics: an introduction. In: *The Ecology of Natural Disturbance and Patch Dynamics*, (eds S.T.A. Pickett, P.S. White), pp. 3–13, Academic Press, Orlando, FL.

Pickett, S.T.A., Collins, S.L., Armesto, J.J. (1987) Models, mechanisms and pathways of succession. *The Botanical Review*, 53, 335–371.

Pokhriyal, P., Chauhan, D.S. & Todaria, N.P. (2012) Effect of altitude and disturbance on structure and species diversity of forest vegetation in a watershed of central Himalaya. *Tropical Ecology*, 53(3), 307–315.

Post, W.M. & Kwon, K.C. (2000) Soil carbon sequestration and land-use change: processes and potential. *Global Change Biology*, 6, 317–328.

Potter, K.N., Tobert, H.A., Johnson, H.B. & Tischler, C.R. (1999) Carbon storage after long-term grass establishment on degraded soils. *Soil Science*, 164, 718–725.

Puget, P., Besnard, E. & Chenu, C. (1996) Une méthode de fractionnement des matières organiques particulaires des sols en fonction de leur location dans les agrégats. *Comptes Rendus de l'Académie des Sciences, Paris, Série II*, 322, 965–972.

Pulford, I. D. & Watson, C. (2003) Phytoremediation of heavy metal-contaminated land trees-a review. *Environmental International*, 29, 529–540.

Pulleman, M.M. & Marinissen, J.C.Y. (2004) Physical protection of mineralizable C in aggregates from long-term pasture and arable soil. *Geoderma*, 120, 273–282.

Purves, D.W. & Pacala, S.W. (2008) Predictive models of forest dynamics. *Science*, 320, 1452–1453.

Raghubanshi, A.S. Srivastava, S.C. Singh, R.S. & Singh, J.S. (1990) Nutrient release in leaf litter. *Nature* 346, 227.

Rapport, D.J. & Whitford, W.G. (1999) How ecosystems respond to stress. *BioScience*, 49(3), 193–203.

Rapport, D.J., Regier, H.A. & Hutchinson, T.C. (1985) Ecosystem behavior under stress. *The American Naturalist*, 125, 617–640.

Raskin, I., Nanda Kumar, P.B.A., Dushenkov, S. & Salt, D.E. (1994) Bioconcentration of heavy metals by plants. *Current Opinion in Biotechnology*, 5, 285–290.

Rebele, F. (1992) Colonization and early succession on anthropogenic soils. *Journal of Vegetation Science*, 3, 201–208.

Rice, C.W. (2002) *Carbon in soil. Geotimes*, January. http://www.geotimes.org/jan02/featurecarbon.html, accessed 9 June 2015.

Richards, J.P. (2002) Sustainable development and the minerals industry. *Society of Economic Geologists Newsletter*, 48, 1–12.

Richards, I.G., Palmer, J.P. & Barratt, P.A. (1993) *The Reclamation of Former Coal Mines and Steelworks*. Elsevier, Amsterdam.

Ritsema, C.J., Dekker, L.W., Nieber, J.L. & Steenhuis, T.S. (1998) Modelling and field evidence of finger formation and finger recurrence in a water repellent sandy soil. *Water Resource Research*, 34, 555–567.

Rives, C.S., Bajwa, M.I. & Liberta, A.E. (1980) Effects of topsoil storage during surface mining on the viability of VA mycorrhiza. *Soil Science*, 129, 253–257.

Roberts, J.A., Daniels, W.L., Bell, J.C. & Burger, J.A. (1988a) Early stages of mine soil genesis as affected by top soiling and organic amendments. *Soil Science Society of America Journal*, 52, 730–738.

Roberts, J.A., Daniels, W.L., Bell, J.C. & Burger, J.A. (1988b) Early stages of mine soil genesis in a Southwest Virginia spoil lithosequence. *Soil Science Society of America Journal*, 52, 716–723.

Robinson, D. (1994) The responses of plants to non-uniform supplies of nutrients. *New Phytologist*, 127, 635–674.

Russell, W.B. & La Roi, G.H. (1986) Natural vegetation and ecology of abandoned coal-mined land, Rocky Mountain foothills, Alberta, Canada. *Canadian Journal of Botany*, 63, 1286–1298.

Sahu, H.B. & Dash, S. (2011) *Land degradation due to mining in India and its mitigation measures. Proceedings of Second International Conference on Environmental Science and Technology*, Singapore, 26–28 February 2011.

Santantonio, D. & Grace, J.C. (1987) Estimating fine-root production and turnover from biomass and decomposition data – a compartment flow model. *Canadian Journal of Forest Research*, 17, 900–908.

Sartz, L. (2010) *Alkaline By-products as Amendments for Remediation of Historic Mine Sites*. Örebro University, Örebro.

Saxena, M. & Asokan, P. (2002) *Timber substitute products from industrial solid wastes*. In: *Proceedings of the 18th National Convention of Environmental Engineers and National Seminar on Solid Waste Management at Bhopal, October 19–20, 2002* (ed R. Srivastava), pp. 192–200.

Scatena, F.N., Moya, S., Estrada, C. & Chinea, J.D. (1996) The first five years in the reorganization of aboveground biomass and nutrient use following hurricane Hugo in the Bisley Experimental Watersheds, Luquillo Experimental Forest, Puerto Rico, *Biotropica*, 28, 424–441.

Schaetzl, R.J., Barrett, L.R. & Winkler, J.A. (1994) Choosing models for soil chronofunctions and fitting them to data. *European Journal of Soil Science*, 45, 219–232.

Schafer, W.M., Nielsen, G.A. & Nettleton, W.D. (1980) Mine soil genesis and morphology in a spoil chronosequence in Montana. *Soil Science Society of America Journal*, 44, 802–808.

Schaff, W., Bens, O., Fischer, A., et al. (2011) Patterns and processes of initial terrestrial ecosystem development. *Journal of Plant Nutrition and Soil Science*, 174, 229–239.

Scharlemann, J.P.W., Tanner E. V.J., Hiederer, R & Kapos, V (2014) Global soil carbon: understanding and managing the largest terrestrial carbon pool, *Carbon Management*, 5, 81–91. http://dx.doi.org/10.4155/cmt.13.77; accessed on 31st Oct, 2015.

Schlesinger, W.H. (1986) Changes in soil carbon storage and associated properties with disturbance and recovery. In: *The Changing Carbon Cycle*, (eds J.R. Trabalka & D.E. Reichie), pp.194–220. Springer-Verlag, New York.

Schlesinger, W.H. (1990) Evidence from chronosequence studies for a low carbon-storage potential of soils. *Nature*, 348, 232–234.

Sencindiver, J.C. & Ammons, J.T. (2000) Minesoil genesis and classification. In: *Reclamation of Drastically Disturbed Lands*, (eds R.I. Barnhisel, R.G. Darmody & W.L. Daniels), pp.194–220. Soil Science Society of America Inc., Madison, WI.

SERI (2002) *The SER Primer on Ecological Restoration*. Society for Ecological Restoration, Science and Policy Working Group 2002. www.ser.org/, accessed 27 June 2015.

Seymour, R.S. & Hunter, M.L., Jr. (1999) Principles of ecological forestry. In: *Maintaining Biodiversity in Forest Ecosystems* (ed. M.L. Hunter Jr.), pp. 22–64. Cambridge University Press, Cambridge, UK.

Sharma, D. & Sunderraj, S.F.W. (2005) Species selection for improving disturbed habitats in western India. *Current Science*, 88(3), 462–467.

Sheoran, V., Sheoran, A.S. & Poonia, P. (2009) Phytomining: a review. *Minerals Engineering*, 2212, 1007–1019.

Sheoran, V., Sheoran, A.S. & Poonia, P. (2010) Soil reclamation of abandoned mine land by revegetation: a review. *International Journal of Soil, Sediment and Water*, 3(2), 1–20.

Shiralipour, A., McConnell, D.B. & Smith, W.H. (1992) Physical and chemical properties of soils as affected by municipal solid waste compost application. *Biomass and Bioenergy*, 3, 261–266.

Shiro, T. & del Moral, R. (1995) Species attributes in early primary succession on volcanoes. *Journal of Vegetation Science*, 6, 517–522.

Shrestha, R.K. & Lal, R. (2009) Offsetting carbon dioxide emissions through mine soil reclamation. *The Encyclopedia of Earth*, November 16.

Shu, W.S., Xia, H.P., Zhang, Z.Q. & Wong, M.H. (2002) Use of vetiver and other three grasses for revegetation of Pb/Zn mine tailings: field experiment. *International Journal of Phytoremediation*, 4(1), 47–57.

Shu, W.S., Ye, Z.H., Zhang, Z.Q., Wong, M.H. & Lan, C.Y. (2003) Restoration of lead and zinc mine tailings in South China. *Acta Ecologica Sinica*, 23(8), 1629–1639.

Shukla, M.K., Lal, R., Underwood, J. & Ebinger, M. (2004) Physical and hydrological characteristics of reclaimed minesoils in Southeastern Ohio. *Soil Science Society of America Journal*, 68, 1352–1359.

Sigler, W.V., Crivii, S. & Zeyer, J. (2002) Bacterial succession in glacial forefield soils characterized by community structure, activity and opportunistic growth dynamics. *Microbial Ecology*, 44, 306–316.

Singh, A., Jha, A.K. & Singh, J.S. (2000) Effect of nutrient enrichment on native tropical trees planted on Singrauli coalfields, India. *Restoration Ecology*, 8(1), 80–86.

Singh, A.N. & Singh, J.S. (2006) Experiments on ecological restoration of coal mine spoil using native trees in a dry tropical environment, India: a synthesis. *New Forests*, 31, 25–39.

Singh, A.N., Raghubanshi, A.S. & Singh, J.S. (2002) Plantations as a tool for mine spoil restoration. Current Science82(12), 1436–1441.

Singh, A.N., Zeng, D.H. & Chen, F.S. (2006) Effect of young woody plantations on carbon and nutrient accretion rates in a redeveloping soil on coalmine spoil in a dry tropical environment, India. *Land Degradation and Development*, 17, 13–21.

Singh, H., Duraiswamy, A., Subramaniam, U. & De, D. (eds.) (1994) Handbook of Environmental Procedures and Guidelines, Ministry of Environment & Forests, Government of India, New Delhi.

Singh, J.S., Raghubanshi, A.S., Singh, R.S. & Srivastava, S.C. (1989) Microbial biomass acts as a source of plant nutrient in dry tropical forest and savanna. Nature338, 499–500.

Singh, L. & Singh, J.S. (1991) Storage and flux of nutrients in a dry tropical forest in India. *Annals of Botany*, 68 (3), 275–284.

Singh, R.S. (1993) Effect of winter fire on primary productivity and nutrient concentration of dry tropical savanna. *Vegetatio (Belgium)*, 106, 63–71.

Singh, R.S. (1994) Changes in soil nutrients following burning of dry tropical savanna. *International Journal of Wildland Fire (USA).* 4 (3), 187–194.

Singh, R.S., Raghubanshi, A.S. & Singh, J.S. (1991a) Nitrogen mineralisation in dry in dry tropical savanna: effects of burning and grazing. *Soil Biology & Biochemistry*, 23, 269–273.

Singh, R.S., Srivastava, S.C. Raghubanshi, A.S. Singh, J.S. & Singh, S.P. (1991b) Microbial C, N and P in dry tropical savanna: effects of burning and grazing. *Journal of Applied Ecology*, 28, 869–878.

Singh, R.S., Chaulya, S.K., Tewary, B.K. & Dhar, B.B. (1996) Restoration of a coal mine overburden dumps - a case study. *Coal International*, 244(2), 80–83.

Singh, R.S., Singh, A.K., Tripathi, N. & Tewary, B.K. (2012) Carbon sequestration in revegetated coal mine wastelands. Coal S&T Project Report, Ministry of

Coal, Government of India, EE/40 GAP/04/EMG/MOC/2008-2009. CIMFR, Dhanbad, India.

Six, J., Feller, C., Denef, K., Ogle, S.M., de Moraes Sa J.C., & Albrecht, A. (2002) Soil organic matter, biota and aggregation in temperate and tropical soils – effects of no-tillage. *Agronomie*, 22, 755–775.

Skjemstad, J.O., Le Feuvre R.P., & Prebble, R.E. (1990) Turnover of soil organic matter under pasture as determined by 13C natural abundance, Aust. J. *Soil Res.*, 28, 267–276.

Source Watch (2011). The Footprint of Coal. http://www.sourcewatch.org/index.php/The_footprint_of_coal#How_much_surface-mined_is_reclaimed.3F; accessed on 1st November, 2015.

Smith, S.E. & Reed, D.J. (1997) *Mycorrhizal Symbiosis*, 2nd edn, pp. 589.Academic Press, London.

Smith, J.A., Schuman, G.E., Depuit, E.J. & Sedbrook, T.A. (1985) Wood residue and fertilizer amendment of bentonite mine spoils: I. Spoil and general vegetation responses. *Journal of Environmental Quality*, 14, 575–580.

Soegaard, H., Jensen, N.O., Boegh, E., Hasager, C.B., Schelde, K. & Thomsen, A. (2003) Carbon dioxide exchange over agricultural landscape using eddy correlation and footprint modelling. *Agricultural and Forest Meteorology*, 114, 153–173.

Song, S.Q., Zhou, X., Wu, H. & Zhou, Y.Z. (2004) Application of municipal garbage compost on revegetation of tin tailings dams. *Rural Eco-Environment*, 20(2), 59–61.

Sopper, W.E. (1992) Reclamation of mineland using municipal sludge. *Advances in Soil Science*, 17, 351–432.

Source Watch (2011). The Footprint of Coal. http://www.sourcewatch.org/index.php/The_footprint_of_coal#How_much_surface-mined_is_reclaimed.3F; accessed on 1st November, 2015.

Šourková, M., Frouz, J., Šantrůčková, H. (2005) Accumulation of carbon, nitrogen and phosphorus during soil formation on alder spoil heaps after brown-coal mining, near Sokolov (Czech Republic). *Geoderma*, 124, 203–214.

Sousa, W.P. (1979) Disturbance in marine intertidal boulder fields: the nonequilibrium maintenance of species diversity. *Ecology*, 60 (6), 1225–1239.

Sousa, W.P. (1984) The role of disturbance in natural communities. *Annual Review of Ecology and Systematics*, 15, 353–391.

Soussana, J.F., Saint-Macary, H. & Chotte, J.L. (2015) Carbon sequestration in soils: the 4 per mil concept. Agriculture and agricultural soils facing climate change and food security challenges: public policies and practices. Paris 2015, Climate Change Conference COP 21-CMP 11, Sept. 16, 2015.

Spaccini, R., Conte, P., Piccolo, A., Haberhauer, G. & Gerzabek, M.H. (2002) Increased soil organic carbon sequestration through hydrophobic protection by humic substances. *Soil Biology & Biochemistry*, 34, 1839–1851.

Sperow, M. (2006) Carbon sequestration potential in reclaimed mine sites in seven east-central states. *Journal of Environmental Quality*, 35, 1428–1438.

Spiecker, H., Hansen, J., Klimo, E., Skovsgaard, J.P., Sterba, H. & von Teuffel, K. (eds.) (2004) *Norway Spruce Conversion – Options and Consequences*, Brill, Boston, MA.

Stanturf, J.A. & Madsen, P. (2002) Restoration concepts for temperate and boreal forests of North America and Western Europe. *Plant Biosystems*, 136, 143.

State of Environment Report (2009) Environmental Information System, Ministry of Environment & Forests, New Delhi, India, www.moef.nic.in/downloads/home/home-SoE-Report-2009.pdf, accessed 27 June 2015.

State of India's Environment – A Citizens Sixth Report (2008) Centre for Science and Environment, New Delhi, India.

Stockdale, E.A., Lampkin, N.H., Hovi, M., et al. (2001) Agronomic and environmental implications of organic farming systems. *Advances in Agronomy*, 70, 261–327.

Struthers, P.H. (1964) Chemical weathering of strip mine soils. *The Ohio Journal of Science*, 64, 125–131.

Subak, S. (2000) Agricultural soil carbon accumulation in North America: considerations for climate policy. *Global Environmental Change*, 10, 185–195.

Suding, K.N., Gross, K.L. & Housman, D.R. (2004) Alternative states and positive feedbacks in restoration ecology, *Trends in Ecology & Evolution*, 19, 46–53.

Suyker, A.E. & Verma, S.B. (2001) Year-round observations of the net ecosystem exchange of carbon dioxide in a native tallgrass prairie. *Global Change Biology*, 7, 279–289.

Sourkova, M., Frouz, J., Fettweis, U., Bens, O., Hutl, R.F., & Santruckova, H. (2005) Soil development and properties of microbial biomass succession in reclaimed post mining sites near Sokolov (Czech Republic) and near Cottbus (Germany). *Geoderma*, 129, 73–80.

Suzina, N.E., Mulyukin, A.L., Kozlova, A.N., (2004) Ultrastructure of resting cells of some non-spore-forming bacteria. *Microbiology*, 73, 435–447.

Swinnen, J., Van Veen, J.A. & Merckx, R. (1995) Carbon fluxes in the rhizosphere of winter wheat and spring barley with conventional vs integrated farming. *Soil Biology & Biochemistry*, 27, 811–820.

Sydney International Investment Group (2014) http://sydneyiig.com/eng/mining.html, accessed 12 June 2015.

Sydnor, M.E.W. & Redente, E.F. (2002) Reclamation of high-elevation, acidic mine waste with organic amendments and topsoil. *Journal of Environmental Quality*, 31, 1528–1537.

Tate, K.R., Scott, N.A., Parshotam, A., et al. (2000) A multi-scale analysis of a terrestrial carbon budget — is New Zealand a source or sink of carbon? *Agriculture, Ecosystems & Environment*, 82, 229–246.

Tate III, R.L. (1985) Microorganisms, ecosystem disturbance and soil-formation processes. In: *Soil Reclamation Processes* (eds R.L. Tate III & D.A. Klein), pp. 1–33. Marcel Dekker, Inc., New York.

Taylor, L.E. (2006) *World Mineral Production 2000–04*, pp. 15–17. British Geological Survey, Nottingham, UK.

Thomas, D. & Jansen, I. (1985) Soil development in coal mine spoils. *Journal of Soil and Water Conservation*, 41, 439–442.

Thompson, P.J., Jansen, I.J. & Hooks, C.L. (1987) Penetrometer resistance and bulk density as parameters for predicting root system performance in mine soils. *Soil Science Society of America Journal*, 51, 1288–1293.

Thompson, I., Mackey, B., McNulty, S. & Mosseler, A. (2009) *Forest resilience, biodiversity, and climate change. A synthesis of the biodiversity/resilience/stability relationship in forest ecosystems. Technical Series No. 43*, pp. 67. Secretariat of the Convention on Biological Diversity, Montreal, QC.

Thompson-Eagle, E.T. & Frankenburger, W.T., Jr. (1992) Bio-remediation of soils contaminated with selenium. In: *Advances in Soil Sciences*, (eds R. Lal, B.A. Stewart), pp 261–309. Springer-Verlag, New York.

Thum, J., Wünsche, M. & Fiedler, H.J. (1992) Rekultivierung im Braunkohlenbergbau der östlichen Bundesländer. *Bodenschutz* 10, 1–38.

Thurman, N.C. & Sencindiver, J. C. (1986) Properties, classification and interpretation of mine soils at two sites in Western Virginia. *Soil Science Society of America Journal*, 50, 181–185.

Tian, G., Granato, T.C., Cox, A.E., Pietz, R.I., Carlson, C.R., Jr. & Abedin, Z. (2009) Soil carbon sequestration resulting from long-term application of biosolids for land reclamation. *Journal of Environmental Quality*, 38, 61–74.

Tietenberg, T. (1998) Disclosure strategies for pollution control. *Environmental and Natural Resources Economics*, 11, 587–602.

Tisdall, J.M. & Oades, J.M. (1982) Organic matter and water-stable aggregates in soils. *Journal of Soil Science*, 33, 141–163.

Torbert, J.L. & Burger, J.A. (2000) Forest land reclamation. In: *Reclamation of Drastically Disturbed Lands*, (eds R.L. Barnhisel, R.G. Darmody & W.L. Daniels), pp. 371–398. American Society of Agronomy, Madison, WI.

Tordoff, G.M., Baker, A.J.M. & Willis, A.J. (2000) Current approaches to the revegetation and reclamation of metalliferous mine wastes. *Chemosphere*, 41, 219–228.

Torri, S.I., Corrêa, R.S. & Renella, G. (2014) Soil carbon sequestration resulting from biosolids application. *Applied and Environmental Soil Science*, 2014, 9.

Tripathi, N. & Singh, R.S. (2008) Ecological restoration of mined-out areas of dry tropical environment, India. *Environmental Monitoring and Assessment*, 146, 325–337.

Tripathi, N. & Singh, R.S. (2011) *Rejuvenation of contaminated mine spoils*. Project Report CIMFR/DST/SD/045/06, submitted to DST, New Delhi. CIMFR, Dhanbad, India.

Tripathi, N., Singh, R.S. & Chaulya, S.K. (2012) Dump stability and soil fertility of a coal mine spoil in Indian dry tropical environment: a long-term study. *Environmental Management*, 50 (4), 695–706.

Tripathi, N., Singh, R.S. & Singh, J.S. (2009) *Impact of post-mining subsidence on nitrogen transformation in southern tropical dry deciduous forest*, India, Environmental Research, 109 (3), 258–266.

Tripathi, N., Singh, R.S. & Nathanail, C.P. (2014) Mine spoil acts as a sink of carbon dioxide in Indian dry tropical environment. *Science of the Total Environment*, 468–469, 1162–1171.

Trippi, M.H. & Tewalt, S.J. (2011) *Geographic Information System (GIS) Representation of Coal-Bearing Areas in India and Bangladesh*. U.S. Department of the Interior, U.S. Geological Survey, Reston, VA.

Truong, P.N.V. (2004) Vetiver grass technology for mine tailings rehabilitation. In: *Ground and Water Bioengineering for Erosion Control and Slope Stabilization*, (eds D. Barker, A. Watson, S. Sompatpanit, B. Northcut & A. Maglinao). Science Publishers Inc., Enfield, NH.

Truong, P.N. & Baker, D. (1998) *Vetiver grass system for environmental protection. Technical Bulletin No. 1998/1*. Pacific Rim Vetiver Network, Office of the Royal Development Projects Board, Bangkok.

Truong, P., Gordon, I. & Baker, D. (1996) *Tolerance of vetiver grass to some adverse soil conditions. Proceedings of the First International Vetiver Conference*, Thailand, October 2003.

Tscherko, D., Hammesfahr, U., Marx, M.C., Kandeler, E. (2004) Shifts in rhizos-phere microbial communities and enzyme activity of *Poa alpina* across an alpine chronosequence. *Soil Biology & Biochemistry*, 36, 1685–1698.

Tuin, B.J.W. & Tels, M. (1991) Continuous treatment of heavy metal contami-nated clay soil by extraction in stirred tanks and counter current column. *Environmental Technology*, 12, 178–190.

Tunstall, B. (2010) *Measuring soil carbon*. http://www.eric.com.au/docs/research/soil/eric_measuring_soil_carbon.pdf, accessed 11 June 2015.

Turner, G.M. & Dale, V.H. (1998) Comparing large infrequent disturbances: what have we learned? *Ecosystems*, 1, 493–496.

Turner, M. S., Stanley, G. A., Smajstrla, C. D. & Clark, A. G. (1994) Physical characteristics of a sandy soil amended with municipal solid waste compost. *Soil and Crop Science Society of Florida Proceedings*, 53, 24–26.

UK Minerals Forum Working Group 2013-14-Future Mineral Scenarios for the UK (2014) *Trends in UK Production of Minerals*. www.mineralsUK.com, accessed 11 June 2015.

UNEP (1997) Industry and environment, *Mining, Facts & Figures*. 20, 1–91.

United States Geological Survey (2005) *2005 Minerals Yearbook: South Africa* (PDF). Retrieved 28 November 2007. http://minerals.usgs.gov/minerals/pubs/country/2005/myb3-2005-sf.pdf, accessed 27 June 2015.

United States Geological Survey (2013) http://minerals.usgs.gov/minerals/pubs/mcs/2013/mcs2013.pdf, accessed 27 June 2015.

Uresk, D.W. & Yamamoto, T. (1986) Growth of forbs, shrubs and trees on bentonite mine spoil under green house conditions. *Journal Range Management*, 39, 113–117.

US Energy Information Administration (2013) *Domestic Uranium Production Report – Annual*. http://i2massociates.com/Downloads/EMDUranium2014 AnnualReport.pdf, accessed 27 June 2015.

USEPA (1995) *Inventory of US Greenhouse Gas Emissions and Sinks, 1990–1994*. US Environmental Protection Agency, Washington, DC.

USEPA (2004) *Inventory of US Greenhouse Gas Emissions and Sinks: 1990–2004*. Office of Air and Radiation, US Environmental Protection Agency, Washington, DC. http://yosemite.epa.gov/oar/globalwarming.nsf/content/Resource CenterPublicationsGHGemissionsUSEmissionsInventory2004.html, accessed 9 June 2015.

Ussiri, D.A.N. & Lal, R. (2005) Carbon sequestration in reclaimed minesoils. *Critical Reviews in Plant Sciences*, 24,151–165.

Utomo, W.H. & Dexter, A.R. (1981) Age hardening of agricultural top soils. *Journal of Soil Science*, 32, 335–350.

Van Du, L. & Truong, P. (2003) *Vetiver grass system for erosion control on drainage and irrigation channels on severe acid sulfate soil in Southern Vietnam. Proceedings of the Third International Vetiver Conference*, China, October 2003.

Verma, T.R. & J.L. Thames (1978) Grading and shaping for erosion control and vegetative establishment in dry regions. pp. 399–409 in *Reclamation of Drastically Disturbed Lands*. F.W. Schaller and P. Sutton, editors. American Society of Agronomy, Crop Science Society of America, Soil Science Society of America, Madison, WI.

Vimmerstedt, J.P., House, M.C., Larson, M.M., Kasile, J.D. & Bishop, B.L. (1989) Nitrogen and carbon accretion on Ohio coal minesoils: influence of soil-forming factors. *Landscape and Urban Planning*, 17, 99–111.

Visser, S., Fujikawa, J., Griffiths, C.L. & Parkinson, D. (1984) Effect of topsoil storage on microbial activity, primary production and decomposition potential. *Plant and Soil*, 82, 41–50.

Vitousek, P.M. & Farrington, H. (1997) Nutrient limitation and soil development: experimental 646 test of a biogeochemical theory. *Biogeochemistry*, 37, 63–75.

Vitousek, P.M. & Reiners, W.A. (1975) Ecosystem succession and nutrient retention: a hypothesis. *BioScience*, 25, 376–381.

Vitousek, P.M. & White, P.S. (1981) Process studies in succession. In: *Forest Succession: Concepts and Application*, (eds D.C. West, H.H. Shugart & D.B. Botkin), pp. 266–276. Springer-Verlag, New York.

Vogl, R.J. (1980) The ecological factors that produce perturbation-dependent ecosystems. In: *The Recovery Process in Damaged Ecosystems*, (ed. Cairns, J., Jr.), pp. 63–94. Ann Arbor Publisher, Ann Arbor, MI.

Vogt, K.A., Vogt, D.J., Brown, S., Tilley, J.P., Edmonds, R.L., Silver, W.L. & Siccama, T.G. (1995) Dynamics of forest floor and soil organic matter accumulation in boreal, temperate and tropical forests. In: *Soil Management and Greenhouse Effects*, (eds R. Lal, J. Kimble, E. Levine & B. A. Stewart), pp. 159–175. CRC Lewis, Boca Raton, FL.

Voorhees, M.E. & Uresk, D.W. (1990) Effects of amendments on chemical properties of bentonite mine spoil. *Soil Science*, 150(4), 663–670.

Waldron, L.J. & Dakessian, S. (1982) Effect of grass, legume and tree roots on soil shearing resistance. *Soil Science Society of America Journal*, 46, 894–899.

Wali, M.K. (1999) Ecological succession and the rehabilitation of disturbed terrestrial ecosystems. *Plant and Soil*, 213 (1–2), 195–220.

Wali, M.K. & Freeman, P.G. (1973) Ecology of some mined areas in North Dakota. In *Some Environmental Aspects of Strip Mining in North Dakota*, (ed. M.K. Wali), pp. 25–47. North Dakota Geological Survey, Grand Forks.

Walker, J.M. (1994) Production use and creative design of sewage sludge biosolids. In: *Sewage Sludge: Land Utilization and the Environment*, (eds C. E. Clapp, W. E. Larson & R. H. Dowdy), pp. 67–74. SSSA, Madison, WI.

Walker, L. (1999) *Ecosystems of disturbed ground*. Ecosystems of the World 16. pp. 868, Elsevier Science, Amsterdam.

Walker, J.L. & Boyer, W.D. (1993) An ecological model and information needs assessment for long leaf pine ecosystem restoration. In: *Silviculture: From the Cradle of Forestry to Ecosystem Management* (comp. L.H. Foley). pp. 138–144, General Technical Report SE-88. U.S. Department of Agriculture, Forest Service, Southeastern Forest Experiment Station, Asheville, NC.

Walker, B.S., Carpenter, J. Anderies, N. (2002) Resilience management in social-ecological systems: A working hypothesis for a participatory approach. *Conservation Ecology*, 6, 14.

Warembourg, F.R., Roumet, C. & Lafont, F. (2003) Difference in rhizosphere carbon-partitioning among plant species of different families. *Plant and Soil*, 256, 347–357.

Waschkies, C. & Hüttl, R.F. (1999) Microbial degradation of geogenic organic C and N in mine spoils. *Plant and Soil*, 213, 221–230.

Washington Department of Ecology Report (2015) Soil Organic Carbon Storage (Sequestration) Principles and Management: Potential Role for Recycled Organic Materials in Agricultural Soils of Washington State. Publication no. 15-07-005.

(https://fortress.wa.gov/ecy/publications/SummaryPages/1507005.html; accessed on 19/10/2015).

WEO (2013) http://www.iea.org/publications/freepublications/publication/weo-2013---executive-summary---english.html, accessed 27 June 2015.

White, P.S. & Pickett, S.T.A. (1985) Natural disturbance and patch dynamics, an introduction. In: *The Ecology of Natural Disturbance and Patch Dynamics*, (eds S.T.A Pickett, P.S. White), pp 3–13. Academic Press, New York.

White, P.S. & Walker, J.L. (1997) Approximating nature's variation: selecting and using reference information in restoration ecology. *Restoration Ecology*, 5, 338–349.

Whittaker, R.H. (1975) *Communities and Ecosystems*, 2nd edn. MacMillan Publishing, New York.

Wild, A. (1987) *Soils and the Environment – An Introduction*, Low Price edn. Cambridge University Press, Cambridge, UK. 287 pp.

Williamson, J.C. & Johnson, D.B. (1991) Microbiology of soils at opencast sites: II. Population transformations occurring following land restoration and the influence of rye grass/fertilizer amendments. *Journal of Soil Science*, 42, 9–16.

Wills, B. (1988) *Mineral Processing Technology*, 4th edn. Pergamon Press, Oxford.

Winjum, J.K. & Schroeder, P.E. (1997) Forest plantations of the world: their extent ecological attributes and carbon storage. *Agricultural and Forest Meteorology*, 84, 153–167.

Wong, M.H. & Luo, Y.M. (2003) Land remediation and ecological restoration of mine land. *Acta Pedologica Sinica*, 40(2), 161–169.

Woodmasee, R.G., Dodd, J.L., Bowman, R.A., Clark, F.E. & Dickinson, C.E. (1978) Nitrogen budget of a shortgrass prairie ecosystem. *Oecologia (Berl.)*, 34, 363–376.

Worley, I.A. (1973) The 'black crust' phenomenon in upper Glacier Bay, Alaska. *Northwest Science*, 47, 20–29.

www.nrsc.gov.in/pdf/P2P_JAN11.pdf, XXXX.

Xu, G., Fan, X & Miller, A.J. (2012) Plant nitrogen assimilation and use efficiency. *Annual Review of Plant Biology*, 63, 5.1–5.30.

Yaalon, D.H. (1975) Conceptual models in pedogenesis: can soil forming functions be solved? *Geoderma*, 14, 189–205.

Yang, B. Shu, W.S., Ye, Z.H., Lan, C.Y. & Wong, M.H. (2003) Growth and metal accumulation in vetiver and two *Sesbania* species on lead/zinc mine tailings. *Chemosphere*, 52, 1593–1600.

Yang, J.-S., Lee, J.Y., Baek, K., Kwon, T.-S. & Choi, J. (2009) Extraction behavior of As, Pb, and Zn from mine tailings with acid and base solutions. *Journal of Hazardous Materials*, 171 (1–3), 443–451.

Yazaki, Y., Mariko, S. & Koizumi, H. (2004) Carbon dynamics and budget in a *Miscanthus sinensis* grassland in Japan. *Ecological Research*, 19, 511–520.

Ying, W. (2008) China shuts more coal power plants; warns on shortage (update 1). *Bloomberg*, 8 June 2008. http://www.bloomberg.com/apps/news?pid=news archive&sid=aiyKnEEj9qZ0&refer=asia, accessed 27 June 2015.

Yoshizawa, S., Tanaka, M. & Shekdar, A.V. (2004) Global trends in waste generation. In: *Recycling, Waste Treatment and Clean Technology*, (eds I Gaballah, B Mishra, R Solozabal, M Tanaka), pp. 1541–1552. TMS Mineral, Metals and Materials Publishers, Spain.

Young, A. (1997) *Agroforestry for Soil Management*, 2nd edn. CAB International, Wallingford.

Young, T.P., Petersen, D.A. & Clary, J.J. (2005) The ecology of restoration: historical links, emerging issues and unexplored realms. *Ecology Letters*, 8, 662–673.

Yunusa, I.A.M., Mele, P.M., Rab, M.A., Schefe, C.R. & Beverly, C.R. (2002) Priming of soil structural and hydrological properties by native woody species, annual crops, and a permanent pasture. *Australian Journal of Soil Research*, 40(2), 207–219.

Zhang, Z.Q., Shu, W.S., Lan, C.Y. & Wong, M.H. (2001) Soil seed bank as an input of seed sources in vegetation of lead/zinc mine tailings. *Restoration Ecology*, 9(4), 1–8.

Index

abandonment, 51, 53, 58, 59
abatement, 52, 53
abiotic, 3, 5, 38, 71
aboveground biomass, 102, 140, 146, 148
accumulation of organic matter, 79
accumulation of soil, 125, 132, 138
acidity, 97, 136
activities, metabolic, 101, 105, 107, 135, 136
afforestation, 7, 52, 53
aggradation, 36, 77
aggregates, 91, 92, 102–104
  stable, 91, 92, 109
aggregate stability, 76, 132, 136
aggregation, 91, 93, 102, 128, 140
agricultural land/soil, 1, 48, 56, 57, 101
agroforestry, 101
alumina, 18
ants, 90, 91, 103
application, 26, 44, 53, 95, 98, 118, 135, 136
assessment, environmental, 8, 66
atmospheric CO_2, 100, 118, 121
Azotobacter Mycorrhiza, 110

bacteria, 5, 90, 103, 105, 135
  aerobic, 133
biomass, 2, 4, 73, 78, 79, 81, 84, 90, 94, 100, 105, 118, 122, 128, 134, 139, 140, 144, 146, 148
  wood, 78

biomass productivity, 101, 127
bio-reclamation, 109
biosolids, 89, 136
biotic, 3, 5
bulk density, 86, 88, 108, 112, 128, 129, 132, 136
burning, 101, 140

capacity, holding, 79, 87, 89, 126, 132, 135
  labile, 78
carbon uptake, 121, 123, 125–127, 129, 131, 133, 135, 137, 139–141, 143–145, 147
cation exchange capacity, 89, 128
CEC *see* cation exchange capacity
ceramics, 24
chalk, 19, 25
chrome, chromite, 12, 14, 18, 19, 22
chromite mining, 18, 19
chromium, 18–19, 23, 68, 98
clay(s), 24, 79, 91, 102, 125, 129, 130
climate change, 2, 121, 122, 145
climax, 36, 74, 77
closure mine, 2, 16, 47, 48, 52, 53, 58, 59, 61, 66, 67, 84, 111, 148
CO_2, 2, 3, 100, 101, 103, 106, 121, 122, 124, 138, 139, 144
coal, 5, 12, 14–15, 21–26, 30, 31, 33–35, 62, 63, 66, 68, 121, 122, 145
  burning, 139, 140
coal mining, 15, 23, 26, 28, 33, 49, 56, 60, 122, 144, 145

Reclamation of Mine-Impacted Land for Ecosystem Recovery, First Edition. Nimisha Tripathi, Raj Shekhar Singh and Colin D. Hills.
© 2016 John Wiley & Sons, Ltd. Published 2016 by John Wiley & Sons, Ltd.